Iron Age to
DARK AGE

1200BC TO AD1000

Published by The Reader's Digest Association Limited
London • New York • Sydney • Montreal

Contents

Introduction

The onset of the Iron Age opened the door on a world that we can still recognise today. Organised in ever more complex societies and increasingly in command of their environment, people learned to master a variety of technologies designed to satisfy their needs and desires. A quick scan of the major inventions of the period serves to suggest the profound changes that were already shaping the world. The emergence of the alphabet helped to diffuse knowledge across the eastern Mediterranean lands and beyond, while the increasing use of written scripts began to preserve knowledge for posterity. Money economies started to replace the barter system, revolutionising commercial activity and possibilities for trade. The Phoenicians, the most active merchants of the era, and Egyptians learned to navigate using compass and maps, pushing back the boundaries of the known world. Great innovators of the era include the traveller Herodotus (5th century BC), the mathematician and engineer Archimedes (3rd century BC) and the architect

Master builders of the Roman Empire
The Pont du Gard aqueduct in southern France is one of the most impressive and enduring legacies of the engineers and builders of ancient Rome. The construction was made possible through the mastery of the arch, a crucial step in the history of architecture. The Pont du Gard was just one of many aqueducts constructed by the Romans to carry fresh water to their cities. They also developed an empire-wide road network, much of which is still with us today.

Vitruvius, who served as military engineer to Julius Caesar (1st century BC).

The geographical focus of innovation was gradually shifting as ageing empires faded and new ones emerged to take their place. Many decisive breakthroughs were made in China, which became a unified empire at this time. The Romans harnessed the knowledge of previous cultures as they rose first to dominate the Mediterranean world and eventually most of Europe. The fall of the Roman Empire would take a whole swathe of knowledge with it, plunging the Continent into the Dark Ages. The Classical heritage that survived, from mathematical and medical treatises to the works of astronomers and philosophers, was preserved and developed by scholars of the Islamic world, which shot to prominence in the late 7th century. Eventually, scholarship was reintroduced to Europe – primarily through Moorish Spain, for a time the leading light of civilisation – where it would flourish in a renewed spirit of enquiry that became known as the Renaissance.

The editors

A bronze Celtic helmet from the 1st century BC. Metal-working techniques were put first and foremost to military use. ▼

◄ Hieratic Egyptian writing marked a step toward the development of the alphabet; this sample is from the 13th century BC.

◄ Kites were invented by the Chinese around the beginning of the 1st millennium BC.

▲ Coins were not always made of metal – this one from 14th-century-BC China is porcelain.

As human beings, we not only know things, we are also consciously aware of possessing knowledge. Through the exercise of our intelligence, we can analyse and come to understand the world we live in, and our ability to learn from

◄ The Parthenon in Athens exemplifies the degree of perfection that Greek architecture had attained by the 5th century BC.

◄ Geometry was in use in Mesopotamia 4,000 years ago, as shown on this tablet defining land boundaries and ownership.

▲ Tunnelling has a long history in both peacetime and war. In the 6th century BC, the Greek engineer Eupalinos constructed a tunnel more than 1km in length. These medieval French sappers are digging to undermine an enemy castle from inside a siege tower.

observation and experience has enabled us to exert mastery over our surroundings. The knowlege we use in daily life is constantly being assessed, updated and corrected, becoming ever more refined and better adapted to our aims. Over the course

▶ Chinese women using a bedwarmer-like device – an ancestor of the iron – to smooth out creases in fabric; the illustration dates from the 12th century.

Double and triple banking of oars enabled Greek biremes and triremes to confront high seas and open up reliable maritime routes in the 5th century BC. ▼

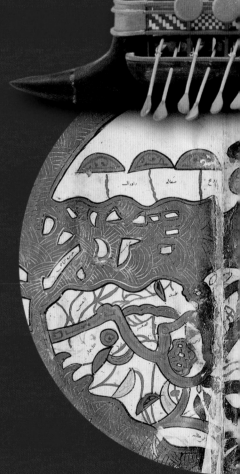

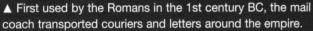

▲ First used by the Romans in the 1st century BC, the mail coach transported couriers and letters around the empire.

of the human saga, tools have extended our capabilities. In time, the techniques employed in using them became objects of study and reflection in themselves, and the more widely tools and techniques were adopted the faster they evolved. Such, at any

▼ Today's lighthouses are successors to the Pharos built at Alexandria on the North African coast in the 3rd century BC.

◄ The first maps were schematic plans, like this 12th-century Arab example, rather than accurate depictions of the lay of the land.

◄ An Arabic manuscript illustrates the concept of gears, invented by Greek engineers in the 3rd century BC.

rate, was the case in the early ages of human history. With the passage of time, direct personal contact proved insufficient to achieve the ends of progress, at least as practised within the confines of the small communities that then made up the social

▲ The horse's bit – the mouthpiece of a bridle – dates from as early as the 4th millennium BC; together with the saddle (invented 5th century BC) and the stirrup (3rd century AD), they gave riders greater control over their mounts.

The concept of the parachute dates back to 2nd-century-AD China, long before Leonardo da Vinci sketched this prototype in his notebooks. ▼

Legend has it that Archimedes was in the bath when he had his 'Eureka moment' and discovered a way of measuring the density of solid bodies. ▼

universe. Things needed to be recorded in some more permanent way. As the world opened up and people's horizons started to widen, humankind entered the age of exchanges – peaceful or otherwise. From that point on individuals, families, clans, tribes and

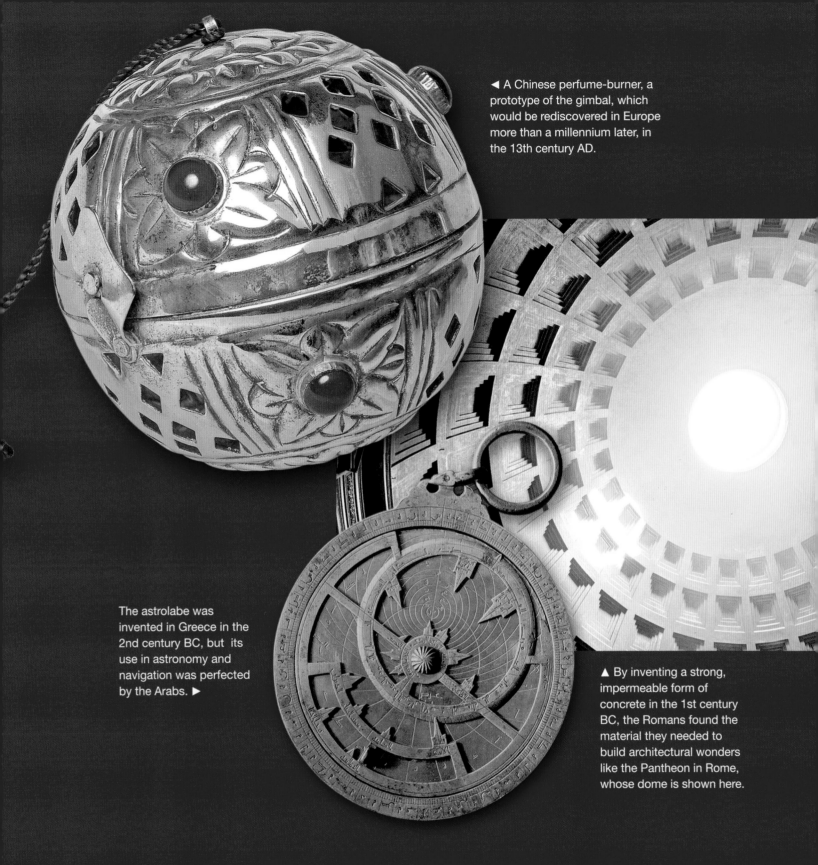

◄ A Chinese perfume-burner, a prototype of the gimbal, which would be rediscovered in Europe more than a millennium later, in the 13th century AD.

The astrolabe was invented in Greece in the 2nd century BC, but its use in astronomy and navigation was perfected by the Arabs. ►

▲ By inventing a strong, impermeable form of concrete in the 1st century BC, the Romans found the material they needed to build architectural wonders like the Pantheon in Rome, whose dome is shown here.

even whole nations began to be less isolated. As partners or rivals, allies or adversaries, they could not avoid rubbing up against one another, sometimes meeting in conflict but more often in fruitful collaboration. The human race had taken a giant step forward.

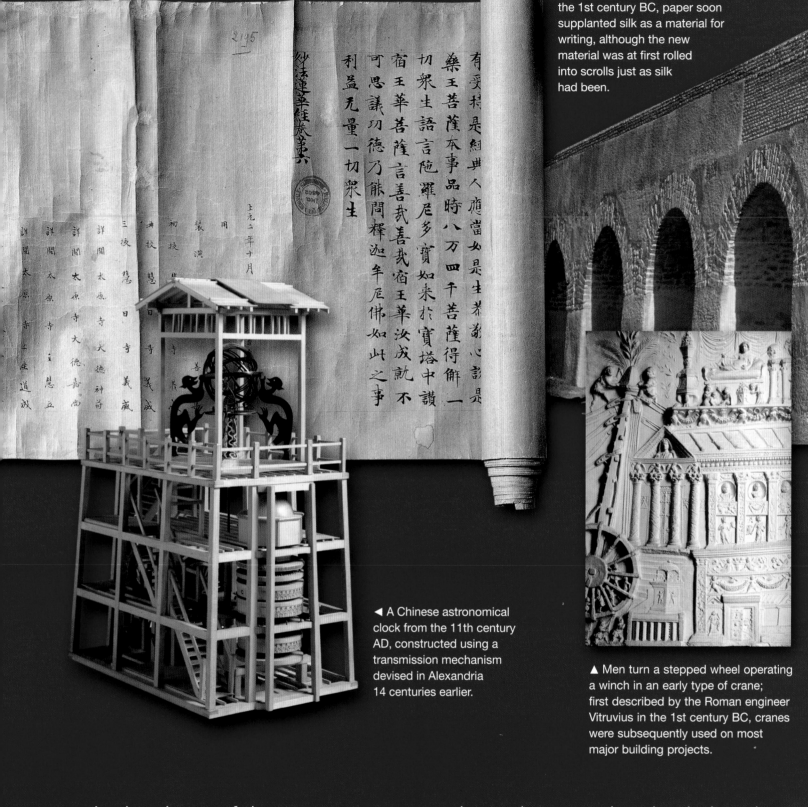

有受持是經典人應當女是生恭敬心設是

藥王菩薩本事品時八萬四千菩薩得解一

切衆生語言陀羅尼多寶如來於寶塔中讃

宿王華菩薩言善哉善哉宿王華汝成就不

可思議功德乃能問釋迦牟尼佛如此之事

利益无量一切衆生

妙法蓮華經卷第六

詳閱太原寺上座道成

詳閱太原寺大德嘉尚

詳閱太原寺大德神符

三校慧日寺義威

初校沙門慧日

辨校沙門善裝

裝潢

用

上元二年十月

2145

◄ Invented by the Chinese in
the 1st century BC, paper soon
supplanted silk as a material for
writing, although the new
material was at first rolled
into scrolls just as silk
had been.

◄ A Chinese astronomical
clock from the 11th century
AD, constructed using a
transmission mechanism
devised in Alexandria
14 centuries earlier.

▲ Men turn a stepped wheel operating
a winch in an early type of crane;
first described by the Roman engineer
Vitruvius in the 1st century BC, cranes
were subsequently used on most
major building projects.

In the dawn of the present era, at a time when empires had already
risen with the dream of unifying the world, our species had moved
into an interconnected environment of technology put to the
service of all. A new imperative now made itself felt: the need to

◄ The Romans brought water to towns by constructing hugely ambitious aqueducts, like this one at Henshir Oudna in Tunisia, dating from the 2nd century AD.

In the interests of administrative and military efficiency, the Romans built a network of paved roads that stretched right across their empire. ▼

◄ The Roman scholar Pliny the Elder is still celebrated as the great scientific populariser of the 1st century AD, although Marcus Terentius Varro had assembled the first encyclopaedia a century earlier.

organise knowledge in a way that would facilitate its transmission and encourage the exchange of ideas. The names of the first great inventors entered the historical record as their discoveries took on a social dimension, transforming communities with their impact.

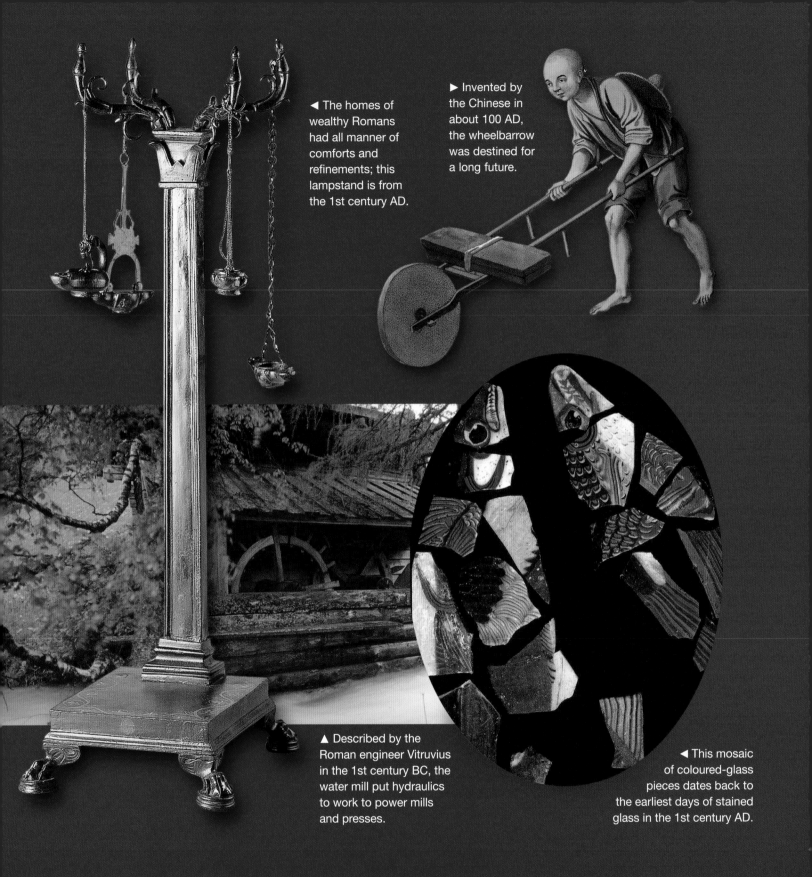

◄ The homes of wealthy Romans had all manner of comforts and refinements; this lampstand is from the 1st century AD.

► Invented by the Chinese in about 100 AD, the wheelbarrow was destined for a long future.

▲ Described by the Roman engineer Vitruvius in the 1st century BC, the water mill put hydraulics to work to power mills and presses.

◄ This mosaic of coloured-glass pieces dates back to the earliest days of stained glass in the 1st century AD.

The first alphabets were chiselled as inscriptions in stone or brushed as phrases onto papyrus. The invention of numerals permitted accounts to be drawn up. Bartering gave way to the money economy. Bridges crossed rivers, helping people to fulfil the

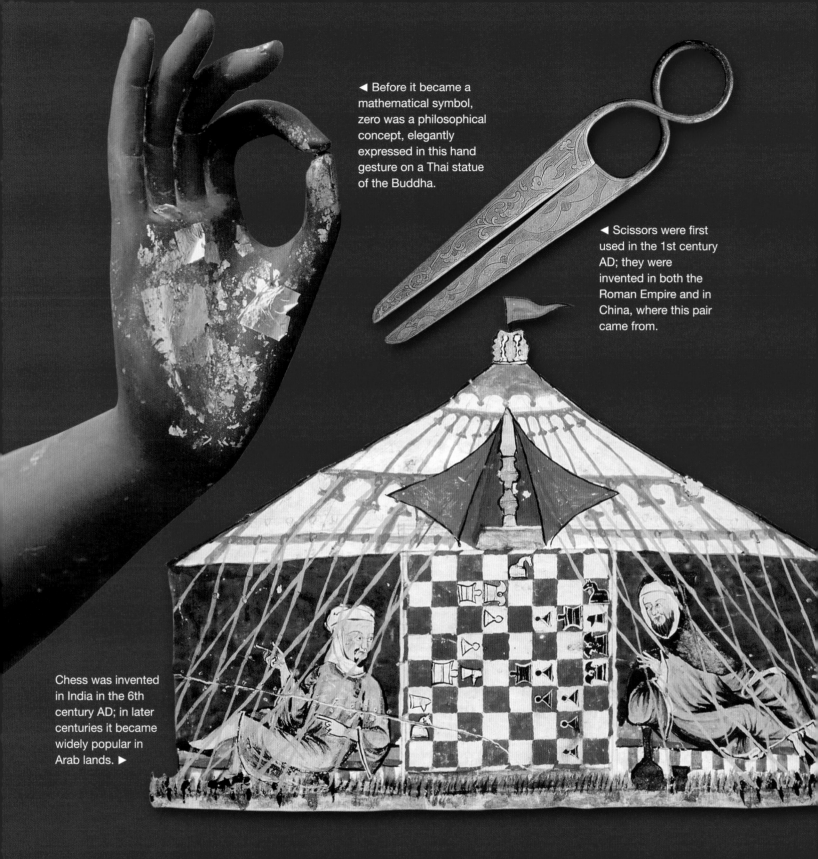

◄ Before it became a mathematical symbol, zero was a philosophical concept, elegantly expressed in this hand gesture on a Thai statue of the Buddha.

◄ Scissors were first used in the 1st century AD; they were invented in both the Roman Empire and in China, where this pair came from.

Chess was invented in India in the 6th century AD; in later centuries it became widely popular in Arab lands. ►

urge to travel and trade. Aqueducts brought water supplies into the heart of cities. The first maps charted frontiers and the path of roads. Hunched in concentration over their worktables, scholars took up the task of gathering knowledge into the compendia that

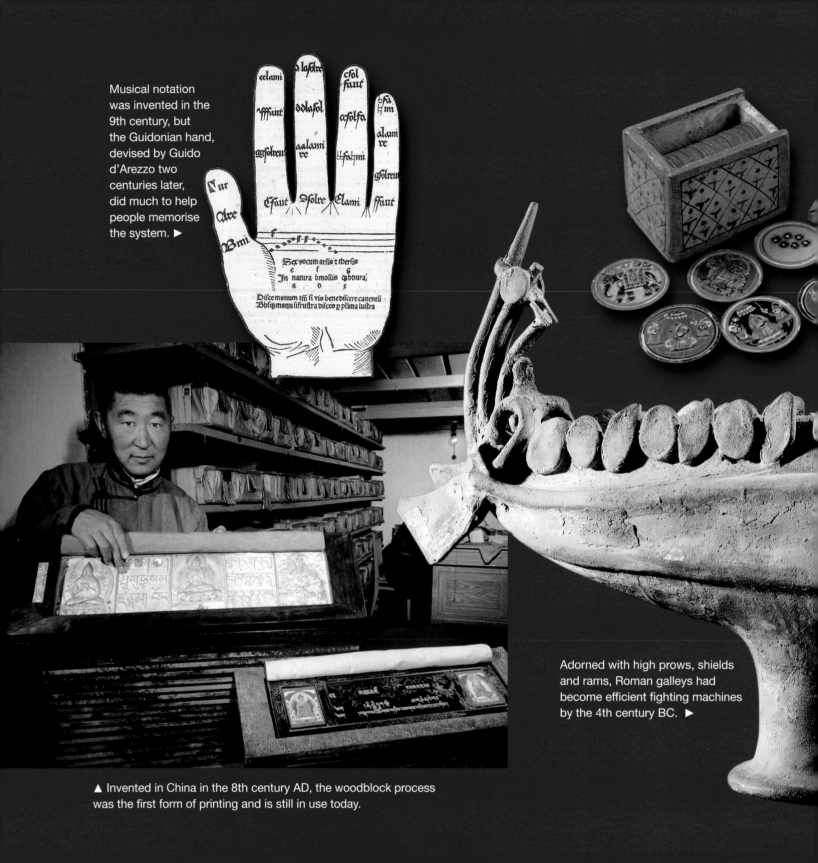

Musical notation was invented in the 9th century, but the Guidonian hand, devised by Guido d'Arezzo two centuries later, did much to help people memorise the system. ▶

Adorned with high prows, shields and rams, Roman galleys had become efficient fighting machines by the 4th century BC. ▶

▲ Invented in China in the 8th century AD, the woodblock process was the first form of printing and is still in use today.

would one day become known as encyclopaedias. Even music, previously only learnt by ear, was now noted down in written form: in future people would be able to learn a song they had not actually heard. Practical concerns would never be far away: every advance

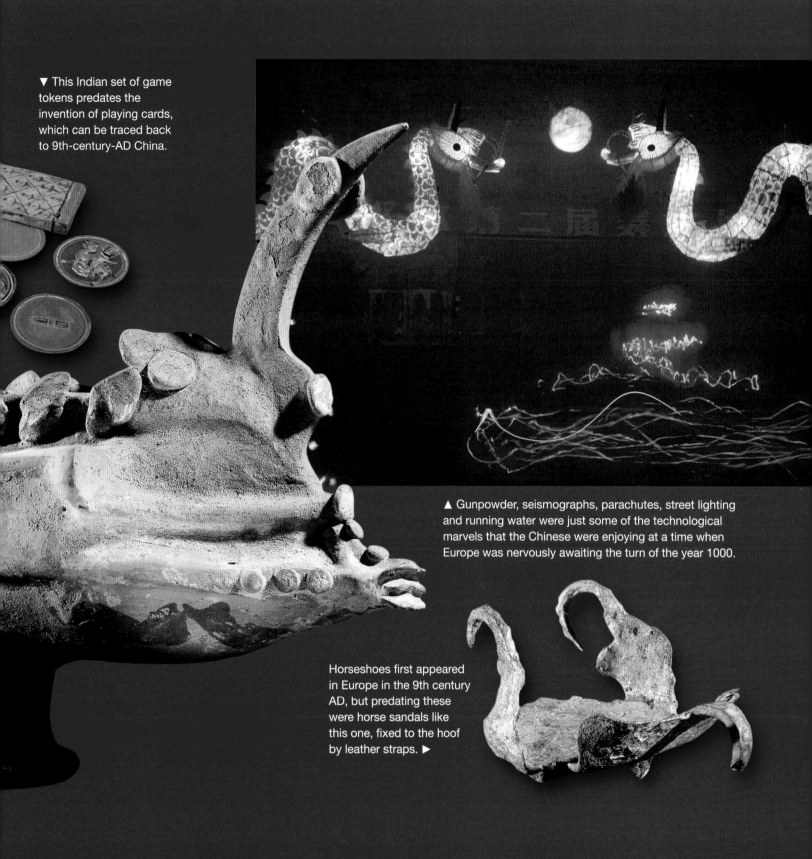

▼ This Indian set of game tokens predates the invention of playing cards, which can be traced back to 9th-century-AD China.

▲ Gunpowder, seismographs, parachutes, street lighting and running water were just some of the technological marvels that the Chinese were enjoying at a time when Europe was nervously awaiting the turn of the year 1000.

Horseshoes first appeared in Europe in the 9th century AD, but predating these were horse sandals like this one, fixed to the hoof by leather straps. ▶

ultimately had to pass the test of greater efficiency – was it a better way of doing things. In its ambition to control the world around it, humankind would need a constantly expanding range of tools, adapted to meet ever more demanding challenges.

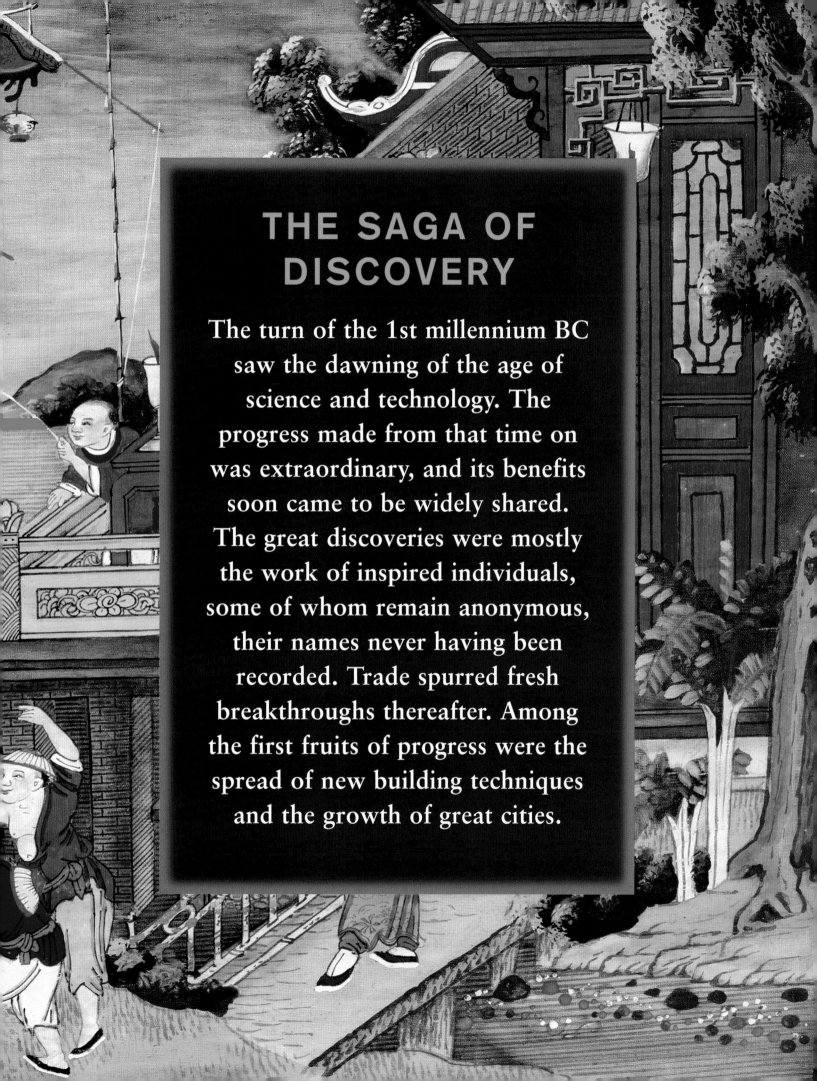

THE SAGA OF DISCOVERY

The turn of the 1st millennium BC saw the dawning of the age of science and technology. The progress made from that time on was extraordinary, and its benefits soon came to be widely shared. The great discoveries were mostly the work of inspired individuals, some of whom remain anonymous, their names never having been recorded. Trade spurred fresh breakthroughs thereafter. Among the first fruits of progress were the spread of new building techniques and the growth of great cities.

IRON – *c*1200 BC

A new age dawns for humanity

First discovered by hunters chipping flakes off meteorites and long considered more valuable than gold, iron changed the destiny of the human race. It proved better than bronze for making weapons, farming implements and a host of other tools. With the breakthrough of ore-smelting, the Iron Age duly got under way and we are living in it to this day.

Military metal
From its very beginning, iron technology was put to military use. The iron arrowheads (background, above) and the decorated iron dagger with its sheath (right) all date from the Hallstatt era in the 6th century BC.

The heavens have been kind to the human race throughout its technological adventure, and rarely more so than in the case of iron. The first specimens of the metal came in an almost pure state from huge meteorites that fell to Earth in Asia. It was in the Far East that iron technology was born, moving out to other parts of the world in successive waves.

The fact that it was subsequently rediscovered in a variety of different venues makes it hard to give universal dates for the start of the Iron Age. China entered it directly from the Stone Age, sometime in the 3rd millennium BC. In Europe the earliest Iron Age sites – those of Hallstatt in Austria and La Tène in Switzerland – date back no further than between 700 and 500 BC.

By the middle of the 3rd millennium the Hittites had brought ironworking technology with them from the plains of Asia to their new home in Anatolia (in today's Turkey), from where it travelled around the eastern Mediterranean lands. A pin and an iron crescent dating back to at least 2500 BC have been found at the Hittite archaeological site at Alacahöyük. The discovery of an iron axe

dated to 1500 BC at Ras Shamra in Syria and an iron dagger with a gold handle among the personal effects in Tutankhamun's tomb in Egypt demonstrate how knowledge and appreciation of the metal gradually spread.

Yet meteorites are rare – so rare that people inevitably started looking for a more abundant and reliable source of iron. As it happened, there was plenty of iron ore to be found in Anatolia, as in the adjoining regions of Armenia, Iran, the Caucasus and the Taurus mountain ranges. At some point around 1200 BC the Hittites learned how to extract and smelt that ore.

HOW THE FIRST METALWORKERS EXPLOITED METEORITES

Thousands of years ago blazing rocks, or meteorites, occasionally fell to Earth from the skies – as they do to this day. Some of these meteorites weighed hundreds of tonnes, others just a few grams. Archaeologists sought long and hard to understand how exactly the world's earliest metalworkers exploited this heaven-sent gift. In fact it turns out that siderites – meteorites largely composed of iron – are relatively rare; in the vast majority of meteorites that reach the Earth, iron makes up only a small part of the total composition.

Mighty meteorite *At 31,000 tonnes this siderite, which fell on Greenland 100,000 years ago, is the second-biggest known.*

In 1894 the American explorer Robert Peary discovered three enormous siderites in Greenland and noted the way in which the local Inuit hunters made use of them. He found that they employed stone hammers to chip away flakes of almost pure iron, which they then inserted into lengths of bone or walrus tusk to make improvised knives. Similarly, the Descubridora meteorite of northern Mexico was found to have the broken blade of a copper chisel wedged within it, indicating that ancient Mayans also sought to break off and use small slivers of the metal.

flames using large bellows with a pottery neck. This began the process of removing impurities from the ore and freeing up the incombustible iron contained within it.

The smelting process took several hours, at the end of which a spongey black substance – the bloom – would be left at the bottom of the furnace. This was transferred to a similar furnace for reheating until softened, then removed with tongs and placed on a primitive anvil, where ashes and other impurities were beaten out with a hammer. This part of the operation required considerable strength, and sparks would fly in all directions.

A new material

The metal ingot that finally emerged had little of the glamour of bronze; it was liable to wear and tear, and if exposed to the damp might even end up covered in unsightly spots of rust. Even so, smiths loved the new metal for its malleability, while its lightness relative to its strength appealed both to warriors using iron weapons and to peasants hefting farming tools. Iron ore also had the advantage of being easy to obtain, coming from surface deposits that were much easier to exploit than those for copper or tin.

If hearsay is to be believed, the Hittites inherited their metal-working skills from an otherwise unknown people whom the Greeks called the Chalybes, said to have lived south of the Caucasus Mountains in a region roughly

The secret of the gods

Blacksmiths were prestigious figures in almost all ancient civilisations, for it was generally believed that only the gods could have taught them to transform rocks and pebbles into gleaming metal. The reality of iron production was altogether more prosaic. To achieve the transmutation the metalworker needed a receptacle filled with charcoal which was placed in a bloomery – a primitive furnace. A number of rust-coloured stones containing iron ore were then thrown into the receptacle and covered with more charcoal. The next step was to light the charcoal and cover the furnace with a clay cone pierced at the base with ventilation holes. An apprentice would fan the

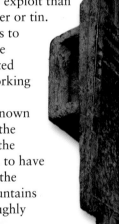

Axe mould
Dating from the 5th or 4th century BC, this Chinese mould for shaping iron axe heads was found with one of the heads alongside it.

Iron ingots
By the 6th century BC the Celts were casting iron into ingots like these (left), which shows how highly the metal was valued.

21

corresponding to modern Armenia. According to Greek tradition, Chalyb smiths were the first to produce iron by smelting sometime around 1500 BC, perhaps drawing on earlier experience of working with ore from meteorites.

It is only possible to speculate as to how the invention came about. One day, perhaps, a blacksmith smelting copper came upon a reddish mineral ore containing not just copper but also iron. Heated between two layers of charcoal, the ore would have partly melted, releasing its copper content but not the iron, which requires much higher temperatures of at least 1500°C that could not have been attained in the primitive furnaces in use at the time. So the smith would have noticed a residue left behind after the copper had been removed, then simply out of curiosity tried hammering it. At this point a seeming miracle would have occurred: under the hammer blows the substance gradually transformed into a grey metal that would prove easy to work.

The blacksmith's tool kit
The ancient tongs and chisel (right) were made 26 centuries ago, yet tools like them are still put to use by skilled craftsmen like the Sudanese blacksmith at work below.

An empire of iron

At some point, the Chalybes' Hittite successors must have had the idea of plunging a red-hot blade, freshly hammered, into water, thereby discovering that tempering metal could increase its strength. Constantly experimenting and innovating, such men were the architects of Hittite power, arming the

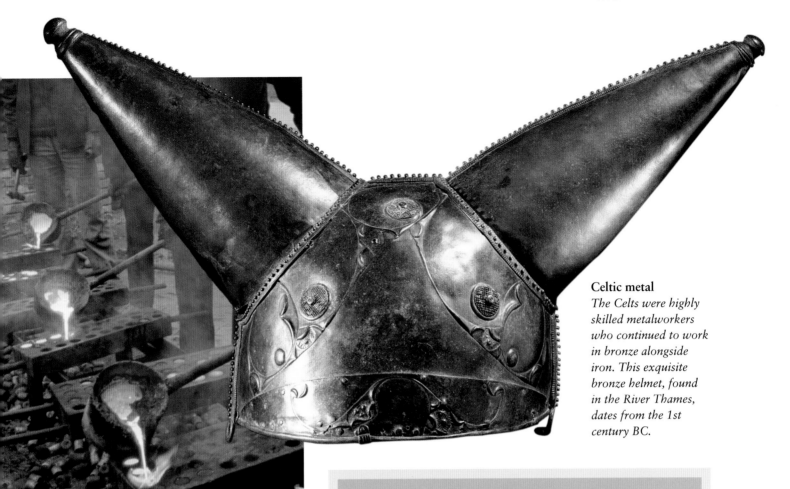

Celtic metal
The Celts were highly skilled metalworkers who continued to work in bronze alongside iron. This exquisite bronze helmet, found in the River Thames, dates from the 1st century BC.

Time-tested ways
In a forge in Sichuan province Chinese blacksmiths pour red-hot molten iron into moulds, in a method little changed from that of earlier Chinese metal-workers since the 1st century AD.

nation's warriors with long swords, protecting them with iron helmets and coats of mail, and encircling the wheels of their war chariots with bands of the same versatile metal. Thus equipped, the warriors were ready to sally forth to conquer new lands.

Hittite smiths were also responsible for the wealth of the nation, for the empire-builders knew how to exploit their discovery by trading in the metal. Thus by land and sea, and through a combination of war and commerce, the Hittites spread the use of iron and probably also the technological knowledge required to produce it.

The spread of metal-working

By the start of the 1st millennium BC smelting was well established in a band of territory stretching from Anatolia through Mesopotamia to the Caucasus Mountains.

FROM HALLSTATT TO LA TÈNE

In 1824, at a salt mine near the village of Hallstatt in Upper Austria, an ancient Celtic cemetery was uncovered. Some tombs contained iron-bladed swords with hilts decorated with iron and amber – evidence of the privileges that metal weaponry brought to the Celtic military elite. The site subsequently gave its name to the earliest Iron Age culture of central Europe. The Hallstatt era saw the emergence of a new type of society, organised in fortified settlements dominated by warrior princes that traded salt and iron around the Mediterranean world. Salt-mining in the area dates back some 3,000 years and was in peak production from the mid 8th to the 6th century BC.

From about 500 BC a second Iron Age culture emerged at La Tène beside Lake Neuchâtel in Switzerland. Discovered in 1853, this site turned up a variety of weapons and ornaments, yet its original purpose remains a mystery: scholars have variously interpreted it as a battlefield, a market and a religious sanctuary.

By the 9th century there were smiths at work not just in the eastern Mediterranean lands but also in Niger, deep inside Africa. The first forges in central Europe and in Italy were established by the middle of the 8th century BC; by 600 BC the technology had reached Spain, Switzerland and France. The evidence suggests that people who had earlier shown the greatest mastery of bronze subsequently became the best ironworkers.

gods of fire, forge and war were lending their support to the switchover from bronze to the new metal. Celtic smiths learned to strengthen strips of wrought iron with carburised metal (iron turned to steel by increasing the carbon content of the metal through further heating). They used this technique – which would one day become known as damascene from its use by the metal-workers of Damascus – to arm their princes with long swords famous for sharpness and suppleness, as well as to provide warriors with lances, shield bosses and the chains that held their sword sheaths to their belts. For farmers they fabricated axes, sickles and adzes. Other objects produced by the Damascus smiths included keys, fire grates and other hearthside utensils that were put to everyday use in people's homes.

Iron in the Roman world

The conquering Romans reaped the benefit of this technological inheritance and put the talents of Gallic, Saxon and British smiths to use. From the 1st century BC heavy ploughs, equipped with iron ploughshares to turn over the soil, made their appearance in the Roman world, heralding better harvests. Working under the patronage of Vulcan, god of fire, the smiths made advances of their own. They introduced wooden fan bellows in place of the leather bags that had earlier sufficed, forcing air through pipes into the base of the furnace and thereby raising the temperature of the fire inside. Some authorities think that they might also have introduced the cementation process, used to convert iron into steel through carburisation, which was already known in the Far East. This involved reheating iron in charcoal mixed with various substances that were thought to harden the metal's surface and make it more resistant – whey was even employed on occasion.

The limits of small-scale production

The fall of the Roman Empire did not halt the progress or iron. The demand for agricultural tools increased as forested land was reclaimed. Seeking to increase the quantity of metal obtained, smiths in the Frankish lands worked out modifications to the bloomeries. From the 8th century the furnaces there were bigger and were lined inside with heat-resistant bricks, which enabled metalworkers to achieve higher smelting temperatures that could also be controlled more reliably.

Iron mask
The silver-coated iron mask (above) is part of an ancient ceremonial helmet. It was found in Syria, which was a Roman province at the time when the mask was made in the 1st century AD.

Celtic smiths

The Celtic peoples were attracted to iron less for its innate qualities than because tin was scarce in their lands. Iron ore was relatively plentiful, and the forests provided the large amounts of wood needed to fuel the furnaces – 4kg of charcoal were required to produce just 1kg of iron. Soon the

Viking smithy
A Norse blacksmith hammers iron while his apprentice works the bellows (above right). This medieval scene was recorded on a wooden medallion in the church at Hylestad in Norway, which was built in the 12th century AD.

AN IRON AGE SITE IN SOUTHERN ENGLAND

At the Kestor archaeological site near Chagford in Devon, archaeologists in the 1950s investigated the remains of an ancient settlement. One of the 27 earth-floored huts in the village contained a smelting furnace in the form of a pit about 45cm across which revealed remnants of iron and charcoal. A second pit nearby may have been used as the hearth for a forge fire, at which the smith reheated the smelted iron before hammering it into its final form.

Yet for all the progress made, medieval ironworkers in Europe ran up against a seemingly insurmountable obstacle. Even the biggest furnaces could only turn out a few kilograms of iron at each firing. And for all the uses to which iron was put, it remained relatively scarce and therefore expensive, as the people who relied on it, from warriors to farm workers, knew to their cost.

A different story in China

In China the occurrence of heat-resistant clays enabled metalworking to develop centuries ahead of the West. Blast furnaces for producing cast iron made their appearance in Europe in the 8th century AD but only became common from 1380; in China they had been in use from the 4th century BC. Just a century later Chinese smiths invented the technique of reheating, a process that involved maintaining the iron at a high temperature over a coal fire, in order to eliminate the sulphur content and make the iron less friable. They also discovered a way of reducing the temperature inside the furnace by throwing vivianite (hydrated iron phosphate) onto the charcoal. Unlike meteor-sourced iron, the cast iron produced was generally not malleable but had to be melted and shaped in moulds. Such methods vastly increased the efficiency of ploughs.

The introduction of cast iron had two unforeseen consequences in China. The first was the production of the first metal containers that could withstand high temperatures – high enough to evaporate brine in order to obtain refined salt. Secondly, the Chinese began to dig deep underground in search of iron ore deposits, and in doing so they discovered natural gas. This was used not just to develop the salt industry but also, as a by-product, to light city streets. None of these advances would have happened had it not been for the progress made in metallurgy.

Iron-working in China took another major step forward in the 1st century AD. The introduction of water-powered fan bellows increased the temperature in blast furnaces to the 1200–1300°C necessary to produce molten cast iron. Thereafter Chinese furnaces turned out thousands of tonnes of the metal each year, providing the raw material for vast quantities of agricultural tools. Europe would not reach a similar level of production until the 14th century.

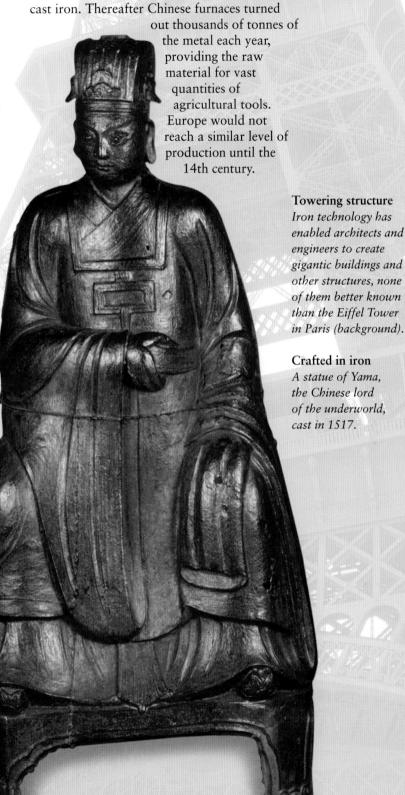

Towering structure
Iron technology has enabled architects and engineers to create gigantic buildings and other structures, none of them better known than the Eiffel Tower in Paris (background).

Crafted in iron
A statue of Yama, the Chinese lord of the underworld, cast in 1517.

A CAST-IRON SKY-SCRAPER

In AD 688, on the orders of the Empress Wu, Chinese workmen constructed the earliest known skyscraper at Louning in Shandong province. The structure took the form of a cast-iron pagoda 90m high. Constructed of superposed elements, the pagoda later collapsed in an earthquake. Undeterred, local ironmasters constructed a new if less ambitious iron pagoda in the year 1105, this one with reaching a height of just 24m.

Symbols that captured a world of meaning

Late in the 2nd millennium BC a new way of writing appeared on the eastern seaboard of the Mediterranean. With 22 letters the Phoenician alphabet was adaptable enough to write down any word. By simplifying reading, it opened a path for the spread of knowledge in many lands.

Message from the grave
An inscription from the tomb of King Ahiram shows that the Phoenician alphabet was in use by the 13th century BC.

A, B, C, D, E . . . Without knowing it, pupils reciting the letters of the alphabet in classrooms today are stirring echoes that travel through the millennia all the way back to ancient Phoenicia, the land of Byblos, now in Lebanon, where an alphabetic form of writing was first introduced. The oldest known alphabetic inscription is an epitaph engraved on the sarcophagus of King Ahiram, a contemporary of Rameses II of Egypt who reigned in the 13th century BC. By the following century the alphabet seems to have been in widespread use in the Byblos region.

A symbol for each sound

The royal epitaph was composed of symbols that corresponded not to a word or syllable, as in earlier writing systems, but to a sound. In the terminology of modern linguistics, they were phonemes. The Phoenician alphabet can therefore be seen as the final step in a long process by which words had first been deconstructed into syllables, then the syllables had been reduced to phonemes. It transformed writing into the transcription of a sequence of sounds, taking meaning from the interrelationship between them.

The 22 symbols that made up the Phoenician alphabet were all consonants, reflecting the fact that the Phoenician people spoke a Semitic tongue in which consonants formed the framework of the language – or, more specifically, the building blocks from which words of the same family were constructed. In similar fashion modern Arabic, another Semitic language, inserts two vowels into the root *ktb*, meaning 'to write', to form *kitab* ('book') or adds a prefix and vowels to make *maktoub* ('it is written').

Making things simple

The Phoenician language was, then, structurally suited to the new system, and Phoenicia itself – a nation of traders – brought together the conditions needed for the alphabet to emerge. Ships from all over the eastern Mediterranean congregated in its ports. The merchants needed a form of writing that would simplify the recording of their business transactions.

Yet the alphabet that emerged was in fact only the final step in an evolutionary process that had got under way in the distant past. Ever since writing made its first appearance in Mesopotamia two millennia earlier, it had been continually evolving. Besides pure pictograms (symbols representing individual objects, as a picture of a rayed disc might indicate the Sun), it had taken in phonetic signs standing either for entire syllables, as in Chinese writing or Mesopotamian cuneiform, or for isolated consonants, as in Egyptian hieroglyphs. Chinese and Egyptian scribes had remained attached to the general principle of ideograms, employing a separate symbol for each concept. But their counterparts at Ugarit (now in Syria) took a different tack. Starting around the 16th century BC, they began to develop the 22-letter consonantal alphabet that finally emerged fully fledged down the coast at Byblos.

It is interesting to speculate on the reasons for the parting of the ways. Maybe Egyptian and Chinese scribes had a vested interest in keeping the act of writing complex, if only to protect their own privileged position as masters of letters. Or was there some deeper cultural divide between the civilisations that,

Phoenician forebear
A clay tablet from the 14th century BC (above) shows the alphabetic symbols developed in Ugarit.

on the one hand, employed pictorial images and symbols and those on the other hand that stuck to abstract signs and concepts?

Mysterious beginnings

Up to the late 19th century, scholars viewed the Phoenician alphabet as having sprung fully fledged from the minds of down-to-earth merchants who needed a simple and rapid form of writing to note down their business transactions. The oldest inscription known at that time dated from the 9th century BC and consisted of a stele describing the campaigns of

Mesha, king of Moab, against the Israelites. Opinion changed in 1906 when the archaeologist Flinders Petrie made a breakthrough discovery on the Sinai Peninsula at Serabit el-Khadem, the site of ancient turquoise mines. There he found inscriptions in a language that marked the transition between Egyptian hieroglyphics and Phoenician. The most remarkable thing about the new writing, since dubbed Protosinaitic, was that the hieroglyphs had lost their symbolic value and instead were being used to

Ancient Egypt's other script
A papyrus fragment (below) displays the hieratic script in daily use in ancient Egypt, from which the alphabetic scripts of the eastern Mediterranean lands ultimately evolved. The text describes the Battle of Kadesh, fought against the Hittites in 1285 BC.

Early inscription
The stele of Mesha (above), king of Moab, dates from 842 BC. By the time it was created, alphabetic scripts were already widely used across the Near East.

27

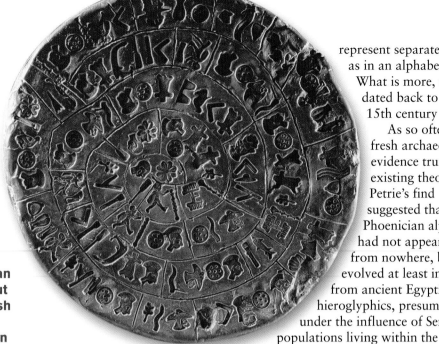

THE MYSTERIOUS PHAISTOS DISC

Taking its name from the ancient palace in southern Crete where it was discovered in 1908, the Phaistos Disc (right) is undeciphered to this day. It contains 122 symbols that do not match any known language. Some scholars have seen a link with a late form of Sumerian, taking the disc to be a votive tablet celebrating the reconquest of the Mesopotamian realms of Sumer and Akkad. But this explanation only raises fresh questions, not least being how did a plaque written in Sumerian find its way to Crete?

represent separate letters, as in an alphabet. What is more, they dated back to the 15th century BC. As so often, fresh archaeological evidence trumped existing theory. Petrie's find suggested that the Phoenician alphabet had not appeared from nowhere, but had evolved at least in part from ancient Egyptian hieroglyphics, presumably under the influence of Semitic populations living within the Egyptian sphere of influence. For although the earliest alphabet was itself largely Semitic, in that 17 of its 22 symbols had Semitic names, it nonetheless also incorporated five symbols whose origins still remain obscure. Scholars do not know whether to trace them to Egypt, Babylon or some entirely different source. What they are agreed on is that Protosinaitic lies at the root of the four great Semitic alphabet families – East, West, South and Central – which between them eventually spawned Aramaic, Hebrew and Greek.

The roots of today's alphabet

The Phoenician alphabet owes its enduring place in history to the wanderlust of the people who employed it and the commercial dealings that took them across the entire Mediterranean basin. In its homeland on the Syrian coast the alphabet survived largely unchanged until the 2nd century AD, but it was also adapted in the surrounding lands to the languages spoken by the Phoenicians' Semitic neighbours. It was improved by the scribes of the kingdom of Damascus, home of the Aramaeans, to form the basis of the Aramaic alphabet, which employed a more convenient, cursive script and doubled up some of the characters to represent vowel sounds.

Aramaic, which was originally used in certain books of the Biblical Old Testament, had many descendants, serving as a progenitor of Arabic as well as of Hebrew. Having used a script derived from Phoenician in the years after 700 BC, the Israelites adopted one directly inspired by Aramaic on their return from Babylonian exile in the 4th century BC and the squared-off, angular Hebrew alphabet survives little changed to this day. Other offshoots of Aramaic included the Pahlavi and

Early Arabic
A 9th-century parchment displays the Kufic script, descended from Old Nabataean.

THE ARABIC LEGACY

The oldest known inscriptions employing an alphabetic Arabic script were found in Syria and have been dated to AD 512 or 513. There is little doubt that the script was modelled on the Phoenician alphabet, although the exact path of transmission is unclear. Since the 5th century BC a South Arabic alphabet comprising 20 consonants had been in use in the southeast of the Arabian Peninsula. In the 2nd century BC the Nabataeans of northern Arabia also used a Phoenician-derived alphabet, composed in their case of 18 characters and 11 diacritical marks, and they too may have played a part in the development of modern Arabic. What is certain is that it was the expansion of Islam out of the Arabian peninsula that carried the script across northern Africa and into East Asia, where it was used not just to transcribe Arab dialects but also Indo-European languages like Persian and Urdu.

Ancient library
*The Thikse
Monastery in
Ladakh in the
Indian Himalayas
preserves an ancient
treasury of written
knowledge (left).*

THE ALPHABET'S PASSAGE TO INDIA

By the 3rd century BC India was in contact both with the eastern Mediterranean lands and with the Arabian Peninsula, and already had two different alphabets employing both consonants and vowels, each one derived from Semitic models. One of them, Brahmi, probably had its roots in Phoenician.

It was to become the basis of the Devanagari script still used today to write down India's sacred language, Sanskrit, as well as the spoken Hindi tongue. The alphabets used in Tibet, Thailand, Cambodia and Laos all subsequently derived from the scripts used in India.

Sogdian scripts, which were both in existence in Iran and Central Asia respectively by the 2nd century BC. They in their turn helped inspire the Turkic and Mongolic alphabets.

Our Phoenician heritage

In the west, the Phoenician alphabet made a first appearance in Greece some 300 years after the land's first, linear scripts had been swept away in the Greek Dark Ages. Subsequently the Greeks adapted the existing alphabet to include vowels. Previously, readers of Aramaic and earlier Phoenician-inspired scripts had needed prior knowledge of the words employed to decipher the writing in a

text. In future, readers could work out the sound of a word simply by mouthing the characters that made it up.

The exact date of the breakthrough remains shrouded in mystery. The new Greek alphabet was in use by the 8th century BC, but there is uncertainty as to where it came from and who originally created it. To convert consonants into vowels they would have required an understanding of both Greek and Phoenician, so it is likely to have happened in one of the places where both languages were spoken: perhaps in Syria, or on the offshore island of

Indian scripts
*A sample from
an 18th-century
illuminated Sanskrit
manuscript (top) and
a 4th-century plaque
engraved in the
Brahmi script
(above).*

29

THE GREEK LEGACY

The Latin alphabet, which like our own had 26 letters, is known to have been in use by the end of the 7th century BC. The Romans inherited it from their Etruscan neighbours, who had in turn acquired it from Greek colonists established on the coasts of Italy. It spread far and wide, carried at first by Rome's own legions and then, after the empire's fall, by European colonisers and missionaries, eventually reaching America, Africa and even Asia, where Portuguese Jesuits employed it to transliterate the Vietnamese language in the 17th century AD.

In similar fashion the Greek alphabet directly inspired the Coptic script used by the Christians of Egypt and east Africa from the 2nd century AD. Byzantine Christianity spread it eastward, into the Armenian and Georgian alphabets and also the Glagolitic alphabet devised by St Cyril to transcribe the Slavic language. This was the precursor of the Cyrillic script still used for writing Russian, Belorussian, Ukrainian, Bulgarian and Serbian.

Poetry and democracy

Such clues would seem to suggest that the first users of the alphabet belonged to an aristocratic class with time on their hands to enjoy banquets and verbal games. For such people the new writing would have been an intellectual challenge. Certainly none of the oldest surviving texts take the form of business papers, chancellery archives, divinations or, indeed, public documents of any kind.

The works of Hesiod and Homer had already been put in writing by about the year 700 BC, so one could almost say that, in marked contrast to most other languages, written Greek had its roots in poetry.

Over the course of the next century, the citizens of the Greek city-states put the new alphabet to civic use to inscribe their laws and honour their gods. By helping spread literacy, the alphabet played an essential part in the development of rhetoric, philosophy and science. By the 5th and 4th centuries BC, elementary schools were standard features of the cities, and a knowledge of reading and writing had already become what it would remain in years to come: the key to knowledge and political participation.

All Greek
A 10th-century manuscript (above) in a Byzantine Greek script.

'I belong to Phidias'
The inscription on the base of this cup (right), made in the 4th century BC, indicates that it belonged to the ancient Greek sculptor, Phidias, a supposition further confirmed by the fact that it was found in his workshop.

Euboea, or even Ionia, the Greek-speaking part of Asia Minor. For all the uncertainty, the first Greeks to use the new alphabet did leave clues to their identities. They were in the habit of marking cups with inscriptions claiming 'I belong to So-and-So', thereby indicating the name of the owner. One individual even used hexameters on a cup to celebrate victory in a dance contest. Another famously parodied *The Iliad* in verses written on a drinking vessel now known as Nestor's Cup, found at an ancient Greek site on the island of Ischia, off Naples.

Kites *c*1000 BC

Most historians think that kites were invented in China, although a minority view holds that they made their first appearance on the Indonesian archipelago. Whatever the case, the first kites were more than just toys. The people of Java and Sumatra were great sailers and used them to ascertain wind direction and speed. As for the Chinese, they saw the lightweight, bamboo-framed, silk-covered wind-catchers as sky-bound vehicles for contacting the gods. More practically, they constructed kites large enough to carry a man and used them to lift tiles up to roofers. They also found various ways of putting kites to military use. One Chinese general is said to have used a kite to measure the distance between his lines and the outer wall of a palace that he was besieging.

There is evidence to suggest that the Mayans invented the kite quite independently on the American continent; the Mayan calendar set aside a day for kite-flying in memory of the dead. Kites did not appear in Europe until around the 16th century, when young and old alike took pleasure in flying rectangular and lozenge-shaped models.

Tools of science

From the 18th century, kites started to attract the attention of Western scientists. In 1749 Alexander Wilson used one to measure atmospheric temperature, while three years later Benjamin Franklin conducted his experiment on the nature of electricity that led to the invention of the lightning rod. Kites also played an important part in paving the way for manned flight and in particular in the development of gliders, notably through the work of the Australian inventor Lawrence Hargrave. Box kites had long been known in the Far East, but in 1893 Hargrave had the idea of joining two together while leaving a gap between, thereby creating the more stable cellular kite. The first biplanes owed much to this development. Kites came back into fashion in the West in the 1970s – they have never gone out of vogue in the Orient – when new materials such as graphite, carbon and glass fibre combined with fresh aerodynamic shapes, some inspired by spinnaker sails, to improve their speed, lightness and manoeuvrability.

Playing with wind
Kites and kite-fliers decorate an 18th-century Chinese porcelain vessel.

Multi-coloured kite
A giant kite in the air at a festival at Ahmedabad in India (left).

American experiment
An illustration (left) purports to show Benjamin Franklin demonstrating the electrical nature of lightning by tying a key to a kite string. In reality, the American inventor flew the kite from a window of his home.

A common currency for commerce

M aking its first appearance in the 7th century BC, money came to perform a threefold function, serving as a means of payment, as a measure of value and as a reserve that could be saved toward some future purpose. Its arrival marked the birth of a new kind of economy – one that was no longer based on barter.

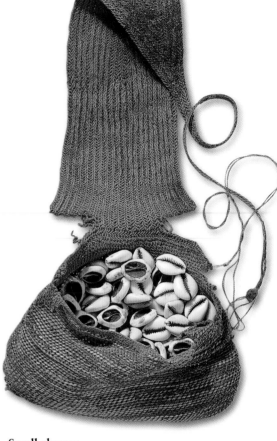

The oldest known coins are an assortment of irregularly shaped discs made of electrum, a natural alloy of gold and silver, that were found at Ephesus in what is now Turkey. Dating from the 7th century BC, they were decorated with lions' heads, the royal emblem of Sardis, which was the capital of the prosperous trading nation of Lydia. It was far from coincidental that money should have first appeared in Asia Minor at this time, for it provided an answer to an apparently complex problem: the need for a generally accepted medium of exchange.

Feathering one's nest
In the South Pacific, coils made from the feathers of rare birds (below) were valuable items of exchange in the days before metal coins were introduced. This example is from the Solomon Islands.

Small change
Cowry shells (left and above) have been used in many countries as a form of money. The woven wallet was made from orchid fibres in New Guinea.

From barter ...

Earlier civilisations had made do with barter, the swapping of one set of goods or services for another deemed to be of equal value. In ancient Egypt, for example, the workers who constructed the Pyramids were paid each week with 200 loaves of bread and five jars of beer; they kept some for their own use and exchanged the rest for other people's produce. If they needed plates to eat from,

for example, they had to find a potter who wanted bread or beer, or who at least had the facilities to store them. If they wanted to buy some scarce commodity whose value was enhanced by its rarity, then they had to find other, more precious means of exchange.

... to the birth of a cash economy

The Lydians were by no means the only ancient people to seek a common currency to replace the barter system. In Polynesia people exchanged fish-hooks; in Melanesia and parts of Africa and South America they used cowry shells. Other African peoples used blocks of salt, while Greeks of the Archaic period (*c*750–480 BC) swapped double-headed axes. The Chinese during the Zhou Dynasty (*c*1045–221 BC) exchanged miniature replicas of everyday objects such as knives and spades. Perhaps the nearest equivalent to the coins of the Lydians were metal ingots used in the 2nd millennium BC in Cappadocia, Assyria and China, whose value depended upon their weight.

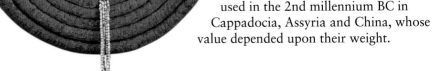

Chinese coin
A painted porcelain coin dating from the 14th century (above).

Rock-solid asset
Doughnut-shaped stone discs were used as money on the Pacific island of Yap in the Carolines group, but they were not portable. The example shown here (above left) is the second-largest on the island.

denomination in relation to the others that were minted. The use of metal coins subsequently spread rapidly across much of the ancient world. In Greece, the island state of Aegina and the cities of Ionia on the Asian mainland (now in Turkey) were the first to adopt the new currency in about 625 BC; the following century, the lawgiver Solon introduced coinage to Athens. Traders in the Persian Empire began using gold and silver coins in their dealings with foreign merchants in about 550 BC, although they continued to make payment in kind for most domestic transactions. In Rome the switchover took place in about 450 BC, when the Law of the Twelve Tables definitively replaced the livestock that had previously been used to settle transactions with the bronze

A new medium with a great future

The Lydian breakthrough came as the last step in a process that had been long in the making. What was new about it was the fact that the currency was issued with the authority of the state, which guaranteed its value by controlling the amount of metal used to produce each coin, as well as by setting the precise worth of each

THE MAKING OF MONEY

The first coins were 'struck' in the literal sense of the word: a strip of metal was stamped in a mould which had a distinctive symbol hollowed out at the bottom. In the early days only one side of the coin bore the distinguishing mark left by the matrix; the other carried a simple hallmark imprinted into the metal with a heavy hammer. At first people were happy to live with some degree of irregularity in coins of the same denomination, but over the centuries technological improvements gradually standardised their weight, size and appearance.

Silver and gold
The gold coin above was struck in the 7th century BC for Croesus, the legendarily rich king of Lydia. The silver coin (left) is a tetradrachma from the kingdom of Macedon and bears the imge of Apollo.

Early bank
In a counting house in Florence in the late 15th century, coins brought in by customers are weighed to determine their precise value.

Promissory note
The ancestors of modern banknotes were, in effect, written promises to pay. This French promissory note (below right) was hand-written on the back of a playing card and pledges 12 livres (pounds).

coinage that the Romans inherited from the Etruscans. In 268 BC, during the war against Pyrrhus, king of Epirus in what is now Greece, the first silver denarii were struck in the temple of the goddess Juno, whose sobriquet Moneta (meaning 'counsellor' or 'adviser') has given us the English word 'money'.

Relative values

Early coins were not entirely satisfactory as a guarantee of value in commercial transactions. Their precise worth depended on the amount of metal they contained and standard weights were hard to maintain given the haphazard methods then used in coin production. Counterfeiting was also relatively easy.

To make matters worse, the financial authorities themselves sometimes had little compunction about advancing their own interests by altering the intrinsic value of the coins minted in their name. There was little to prevent a king from issuing underweight coins or reducing the precious-metal content while leaving the coins' face value unchanged. The fact was that no currency had an absolute value that could always be counted on; instead, confidence played a vital part, and each state's money was only ever worth what the people who used it thought it was worth.

Money and power

The money in circulation in a nation inevitably reflects the way in which that nation's society is organised, as well as its political and economic situation. So it was that, between the 7th and 3rd centuries BC, almost 2,000 separate currencies were minted in Greece, which was then made up of a multitude of separate kingdoms and city-states. In contrast, a powerful, centralised empire was able to impose its money over a vast geographical area. The most spectacular example was the Roman solidus. Starting life as a gold coin issued by the Emperor Constantine in the early 4th century AD, the solidus spread across the Western world during the next 400 years. It subsequently served as the model for the Arab dinar and for the besants minted in the name of the Byzantine emperors.

Intrinsic value or government fiat?

From the 13th century on, the rebirth of trade breathed new life into Europe's currencies. Gold, the most precious metal of all, became a hugely sought-after commodity, and the search for new gold supplies was a major inspiration behind the great voyages of exploration that set off in the 15th century.

The demands of trade also encouraged the invention of fresh financial instruments: bills of exchange were introduced in the 15th century, followed by the banknote, first adopted in Europe in the 17th century but known to the Chinese much earlier – Marco Polo commented on China's paper money after travelling to the Orient in the 13th century. Yet coins proved durable. Hard-wearing, easy to carry and stamped with some distinguishing symbol to guarantee their authenticity, they remain the most widely used of currencies around the world to this day.

THE FIRST BANKNOTES

The idea of creating a symbolic currency whose intrinsic value was set by the issuer goes back to 140 BC, when the Chinese Emperor Wu put squares of deer skin bearing the imperial seal into circulation. Sweden issued Europe's first banknotes in AD 1661. In 1685 Jacques de Meulles, governor-general of the French possessions in North America, ran short of funds and paid his soldiers with playing cards with a cash value written on the back. The watermark, designed to prevent forgery, was introduced by the Bank of England in 1697.

The grammar of the builders' art

Ever since the building of the first cities, a succession of civilisations – Sumerian, Egyptian, Babylonian, Indian, Chinese, Mayan – had constructed not just streets and houses but also grandiose monuments. From the 5th century BC, the builders of Athens gave the world the rules of an architecture designed on a human scale.

Whether by luck or design, the twin roots of the word 'architecture' – *arkhe*, meaning 'principal', and *tektonikos*, 'builder' – are both Greek. The derivation is fitting, since Classical Greek architecture has come to be regarded through the centuries, particularly in post-Renaissance times, as the pinnacle of the art of constructing harmonious buildings.

It took many centuries for the Classical style to come to fruition. From about 1200 BC, successive waves of invaders swept across Greece: Ionians, Achaeans and Dorians each brought with them their own architectural models. Arriving perhaps from Crete or from Mycenae (the question is unresolved), the Dorians favoured a majestic style, influenced by the Orient, with colour as a distinctive feature. At Mycenae itself, where stone replaced wood at an early date, builders cultivated an image of brute power. The blocks used there were so massive yet so cunningly fitted together that the legend spread that the palace walls had been constructed by the Cyclopes, a mythological race of one-eyed giants.

Ancient palace
The palace of Tiryns, with its Cyclopean tunnels (right), was constructed on the Peloponnese peninsula of Greece in the 13th century BC.

Temple of Poseidon
Built in 460 BC at Paestum in Italy, this temple (below) is now thought by some to have been dedicated to the goddess Hera, rather than Poseidon. Whoever the dedicatee, the temple is one of the most celebrated examples of Classical Greek architecture outside Greece itself.

The demands of the elite

Stable and prosperous, the Greek city-states gave birth to intellectual elites of scholars, poets, philosophers and men of letters. These individuals of refined tastes soon turned their attention to architecture and particularly to temples, dedicated to and regarded as dwelling places of the gods. Before long voices were raised complaining that the archaic Doric style, as later exemplified by the Temple of Poseidon at Paestum in Italy, was too squat; the columns were too thick and too close together, making it difficult for processions to pass between them. Such criticisms paved the way for a new lightness of touch. In future the columns would be slimmer and set further apart; for aesthetic as well as practical reasons, architects eventually settled on a ratio of 1.66 times the column diameter for the gaps between them. Sometimes monolithic, at others composed of stone segments clamped by iron bands, the columns had to support pediments and coffered roofs made of stone slabs resting on wooden or marble beams.

military strategist as well as the city's most prominent statesman. He entrusted the rebuilding of the Acropolis to the architects Iktinos and Kallikrates under the direction of the great sculptor Phidias. Their task was to reconstruct the Propylaea (monumental gateway), the Chalkotheke ('Bronze Store'), and the Parthenon, a temple of Athene, as well as a statue and a smaller temple dedicated to the same goddess, the city's divine patron and namesake. Elsewhere in the capital other major developments also got under way.

Jewel of the Acropolis

The most admired of all the structures in Athens is the Parthenon. Majestic without being overpowering, this Doric temple built entirely of marble is peripteral – that is, surrounded on all four sides by columns. The columns are separated on a 1:1.66 ratio and many details reveal the skill of the temple's builders. The final course of the foundation, for example, is not completely flat but curves gently up at the edges, so that the columns slope very slightly inward. The visual effect of this is to correct an optical illusion that would otherwise make the horizontal lines of the building appear slightly curved. For similar reasons, the columns bulge slightly in the middle, an effect most marked in the corner pillars. Such artifices contribute to the perfect harmony of the building as seen by the naked eye – the ultimate arbiter of beauty for the Greeks, who believed, in the words of the philosopher Protagoras, that 'man is the measure of all things'.

Designs for city living

Greek civilisation was essentially urban, and city life focused on the agora or marketplace, the centre of civic life. In Classical times these open spaces were built to a regular plan, decorated with statues and bordered by colonnades sometimes two storeys high, with galleries for business premises and public offices. Stadiums, theatres, gymnasia and *palaestra* (wrestling schools) were hubs of social life. In the construction of the theatres, which were often built on slopes to take advantage of the downward lie of the land, Greek architects revealed their mastery of acoustics. All these monumental structures relied on slave labour for their construction.

Private homes

The houses of Bronze Age Mycenae already exhibited the fundamental characteristics of Greek domestic architecture. They were two or more storeys high, organised around square or

Universally admired
Almost from its inception, the Parthenon in Athens marked the summit of Classical Greek architecture. It is the chief surviving monument of the city's Periclean era.

The flowering of Athens

Through a process of gradual refinement, the Greeks established the laws of proportion in architecture. The Classical style married strength and pomp on the one hand with finesse and sobriety on the other. The builders' art first reached its apogee in Athens, then spread out across the rest of the nation.

In the 5th century BC the city of Athens and the confederation of city-states it headed reached a crucial point in their development. The Peace of Callias, signed in 449 BC, finally brought an end to the destructive Persian Wars, which had left in ruins the monuments topping the Acropolis, as the hill dominating Athens was known. Over the ensuing peaceful decades the Athenians had time on their hands to reconstruct and to improve their city.

The man who supervised the rebuilding was Pericles (*c*495–429 BC), a leading general and

A MATHEMATICAL KEY TO HARMONY

The Greek mathematician Pythagoras (c570–490 BC) is sometimes credited with discovering the so-called 'golden ratio' of 1:1.66 by which two structures of unequal size are most harmoniously ranged against one another. According to the rules, a 'perfect' rectangle would be one in which the longer sides were 1.66 times the length of the shorter ones. In fact, the ratio was employed not just in many of the buildings of ancient Greece but also in the construction of the Great Pyramid of Giza in Egypt. The architects of Renaissance Europe rediscovered the golden ratio by studying ancient models and made great use of it in their own designs.

THE ART OF ACOUSTICS

Drawing on Pythagorean experiments with the diffusion and refraction of sound, the sculptor and architect Polykleitos the Younger showed a complete mastery of acoustics in designing the theatre at Epidaurus on the Peloponnese peninsula in the 4th century BC. Even the spectators farthest from the stage could hear every whisper or rustle of paper in the orchestra, as the circular space where the chorus sang and danced was known. Modern visitors to the site rarely pass up the chance to test out this claim.

rectangular interior courtyards, and simply constructed of brick and wood. The Classical era provoked a taste for decoration in the form of frescoes, wooden surrounds, terracotta ornaments, ceramic or mosaic panels, and sometimes colonnades around the central court. In many cases the basic groundplan was adapted to include two separate courtyards, one giving onto the men's apartments and the other serving the women of the household.

Perfect acoustics
Built in the 4th century BC, the theatre at Epidaurus in the Peloponnese (above) is proof of the technical proficiency of the finest Greek architects.

The Mediterranean heritage

The death of Alexander the Great in 323 BC marked the transition from the Classical to the Hellenistic era of Greece, in the course of which Greek civilisation spread out across the Mediterranean Basin, notably to Sicily and Magna Graecia, the coastal area of southern Italy colonised by Greek settlers. A taste for Corinthian columns with elaborately scrolled capitals took root there, influencing subsequent Roman architecture. Under Eastern influences from Egypt and Mesopotamia, vaulted roofs came back into use, having gone out of fashion in Classical times. Greek house design similarly survived; adopted by the Romans, it became the model favoured in the Mediterranean lands to the present day.

The Erechtheum
Combining grace with majesty, caryatid statues take the place of columns in this temple dedicated to the goddess Athena on the Acropolis in Athens (right).

The art of simplifying the complicated

For the ancient Greeks, three disciplines held the key to knowledge: philosophy, mathematics and geometry. Originally inspired, perhaps, by the study of astronomy and of algorithms, geometry's elucidations of abstract space radically altered existing ways of thought through their universal applicability and irrefutable nature.

Practical science
A 4,000-year-old cuneiform tablet from Mesopotamia employs geometric designs to settle a question of land ownership.

Papyrus fragment
Dating back at least 3,500 years, the Rhind Papyrus (right) is named for the antiquarian who first purchased it. Set out on the papyrus is a series of geometrical problems with their solutions alongside, in a form still familiar in modern schoolbooks.

Greek geometer
This relief of Euclid, the great Greek mathematician, was sculpted in 1334 by the Florentine Andrea Pisano, working from an antique bust.

The word 'geometry' derives from a Greek term meaning 'to measure the land'. In effect, the science first started to take shape when humankind adopted a settled lifestyle, and matured as people began to buy and sell estates and to pay taxes on individual holdings. The first known geometrical drawings date back to Babylonian land registries of the 2nd millennium BC. Confronted with questions concerning the exact size of landholdings with irregular boundaries, the first geometricians learned to convert complex shapes into simple squares and circles whose surface area could be

precisely calculated. Operating by trial and error, they worked out an approximate value for pi (π), the ratio of the circumference of a circle to its diameter – they settled on a value of 3, while the actual value (to four decimal points) is 3.1416. The Babylonians were also the first to divide the circle into 360 degrees.

Ancient Egyptian savants also made a contribution. The Rhind Papyrus, dating to the 17th or 16th centuries BC, reveals how to calculate the area not just of rectangles but also of more complex shapes, such as isosceles triangles and trapeziums. The Egyptians even improved on the Babylonians by coming up with a closer approximation to the true value of pi, producing a figure of 3.16.

Greek astronomy and logic

From such beginnings the Greeks raised geometry to a new level of sophistication as a scientific discipline capable of resolving problems of ever greater complexity. The five theorems of Thales of Miletus (c625–547 BC) opened up the science of astronomy by enabling scholars to calculate the distance of far-off heavenly bodies. Thales himself used his fifth theorem – which stated that it is possible to describe a triangle from the length of the base and the angles it makes with the other two lines – to work out from the seashore how far passing boats were out to sea. His successor Pythagoras (c570–490 BC) was credited with the discovery that the square of the hypoteneuse of a right-angled triangle is equal to the sum of the squares of the other two sides.

For the ancient Greeks, geometry was inseparable from philosophy. It was natural, then, for Athenian sophists to pose the three best-known geometrical praradoxes as brain-teasers to alert geometricians to the limitations of their knowledge. The puzzels concerned the squaring of circles, trisecting of angles and doubling of cubes. From the time of Euclid in Alexandria in the 3rd century BC right up until the 19th century, the three questions remained unresolved, eventually contributing to the birth of a non-Euclidian geometry.

For all that, science's debt to Euclid is huge. The 13 books of his *Elements* were the foundation of all mathematical thought up to the present day, even for non-Euclidians. With the help of 35 definitions, ten axioms and five postulates, he summed up and put in order all the knowledge that had come down to his time. He also helped to found the science of logic, treating it as a discipline bringing together geometry and arithmetic.

Mathematicians at work
Early geometricians demonstrate their art in the 13th century (above) and the 15th century (right).

Fresh complexities

The Greek contribution continued after Euclid's day. Archimedes (287–212 BC) discovered how to calculate the area of circles, spheres and cylinders, and put geometry at the service of mechanics, as his famous remark about levers suggests: 'Give me a place to stand on, and I will move the Earth'. By his day Greek geometry was nearing the end of its development, although his younger contemporary Apollonius of Perga (c262–180 BC) did important work on conic sections.

Much of the Classical Greek heritage was subsequently preserved by the Arabs and the Moors in Spain, who in turn passed it on to the scholars of medieval Europe. It was only with the Renaissance, however, that the sum of geometrical knowledge began to increase once more. By that time the science that had begun by seeking to simplify reality had become dizzyingly complex, testing the finest intellects with its challenges.

THE ARROW THAT NEVER REACHES ITS MARK

One of the best-known geometrical paradoxes was first proposed by Zeno of Elea in the 5th century BC. He suggested that an arrow fired by a bowman should in theory never reach its target because its trajectory, like any line, is composed of an infinite number of separate points, so that traversing them should be similarly endless. Twenty-five centuries later Einstein's theory of general relativity would go some way to resolving the problem.

Tunnels c530 BC

In about 530 BC Polycrates, the tyrant of Samos, commissioned the architect Eupalinos of Megara to construct a tunnel through Mount Kastro on the Greek island that he ruled. As with many of the other great public works of antiquity, the aim was to improve the water supply, in this case to the island's capital.

Even in Polycrates's day there was nothing new about digging underground aqueducts; one had been constructed at Gezer in what is now Israel around the turn of the 2nd millennium BC. Scholars today are still uncertain how these early engineers detected the subterranean aquifers that they tapped.

The first surveyors

The Eupalinian Aqueduct stretched for 1,036m and has the distinction of being the first underground aqueduct constructed with the aid of scientific techniques. Archaeological investigation has established that the work was carried out from both ends at once. Two teams of workers each cleared about 15cm a day, using pick-axes to extract some 5,000 cubic metres of rock. It remains unclear how Eupalinos settled on the entry and exit points for the tunnel, or how he ensured that the twin excavations met; evidence on the ground suggests that even if the orientation was not quite perfect, the two sets of workers managed at least to emerge on the same level.

Sappers strike
Tunnelling played a part in warfare from early times. This manuscript illumination of 1326 shows French sappers (military excavators) working to undermine a castle wall from the shelter of a siege tower (right).

Eupalinos's aqueduct
This tunnel on the Greek island of Samos was constructed to transport water underground more than 2,500 years ago.

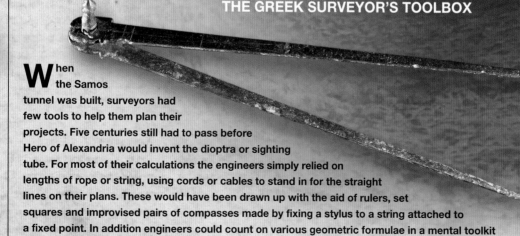

THE GREEK SURVEYOR'S TOOLBOX

When the Samos tunnel was built, surveyors had few tools to help them plan their projects. Five centuries still had to pass before Hero of Alexandria would invent the dioptra or sighting tube. For most of their calculations the engineers simply relied on lengths of rope or string, using cords or cables to stand in for the straight lines on their plans. These would have been drawn up with the aid of rulers, set squares and improvised pairs of compasses made by fixing a stylus to a string attached to a fixed point. In addition engineers could count on various geometric formulae in a mental toolkit that enabled them to work out distances and surface areas from the known dimensions of a site.

Ancient tool
Bronze compasses (left) that were used in the Roman Empire in the 1st century AD.

UNDERGROUND CANALS

Employed in Armenia from the 8th century BC, qanats were widely used in the Middle East to tap underground aquifers. Essentially, qanats are gently sloping tunnels that use gravity to channel water resources downstream in a way that resisted evaporation. To create the tunnels, vertical shafts were drilled at intervals of 30 or 40m; these served not just as access points for the tunnellers but also provided ventilation and somewhere to extract the excavated soil. The shafts also served to mark out the tunnel's route, acting as signposts for maintenance workers. For once a qanat was in operation sediment tended to accumulate in the tunnel as a consequence of the shallow gradient of the incline, and the shafts gave access to clean this out.

Qanats are still in use over a vast region stretching from northern Iran and Iraq to Turkestan. In Afghanistan they are called *karez;* the equivalent term in the Maghreb and the Sahara is *foggara.*

Iranian qanat *The tunnels and the shafts that access qanats form wide-ranging irrigation networks that help to bring otherwise arid regions into cultivation.*

Thames tunnel
Built in the first decade of the 20th century, the Rotherhithe Tunnel (below) was the first road tunnel under the Thames. It took four years to construct and opened in 1908.

The birth of civil engineering

Roman engineers improved drilling techniques, splitting or shattering rock faces by exposing them alternately to extreme heat and cold. In the 9th century AD the Chinese invented gunpowder, which was used exclusively for military purposes when it first reached the West, but many centuries later was put to use in engineering projects. One of the first recorded instances of the civil employment of gunpowder was in France in about 1670, when it was used to blast a tunnel during the construction of the Canal du Midi. It was also used to blast the Standedge Tunnel in the Pennines, the longest, highest and deepest canal tunnel in Britain. This was completed in 1811, shortly before gunpowder was abandoned in favour of more stable and efficient dynamite.

New times, new techniques

The coming of the railway in the 19th century increased the need for tunnels, not least because of the requirement to keep rail gradients below about 1 in 30. The mechanisation of drilling techniques at this time vastly aided the process. The invention of the pneumatic drill, first used in 1860 for mining operations, proved a particular boon; when the Mount Cenis railway tunnel between France and Italy was constructed in 1871, workers were able to advance at a rate of 2.5m a day, ten times what they could have previously managed.

Tunnelling received a further boost with the arrival of the motor age, as a fresh need arose for quick road links between adjoining valleys. The basic techniques employed have barely changed in more than a century, although materials and methods have been improved to make operations safer, faster and more efficient. Modern tunnel-boring machines equipped with integrated shields can clear as much as 40m of tunnel a day.

The world's first long-distance communication networks

Before it became a mail delivery system serving both the private and public sectors, the post was an arm of the state designed for the convenience of people in power. It would remain so until the technology needed to support long-distance networks finally became available.

Even into the 20th century Australian aborigines sent messages in the form of notched sticks, using a secret language that only the bearer could fully interpret. Similar forms of communication that did not require a written language probably date back to the very earliest times. Other examples include the quipu, knotted cords that the Incas and other Andean peoples used to record information.

By the start of the 2nd millennium BC, couriers were travelling the length and breadth of Egypt carrying official messages written on papyrus. Yet organised postal services can only truly be said to have got under way in the 6th century BC. An essential precondition for their development was the growth of centralised states that needed to send regular dispatches out to the provinces and receive communiqués back in return.

Mail coach
A bas-relief preserved in the masonry of the Church of Maria Saal in Carinthia, Austria, shows a Roman mail coach of the 1st century BC (below).

A Persian invention

The Greek historians Herodotus and Xenophon both described the system put in place in about the year 520 BC by Cyrus the Great, ruler of Persia and creator of an empire that under his successors would eventually stretch from the River Indus to the shores of the Mediterranean. In time, a network of imperial roads came to criss-cross this vast domain, which was divided for administrative purposes into 20 separate satrapies, each one governed jointly by a civil administrator and a military commander. Along the length of these roads were

From the top
The message on this clay tablet was sent from Hammurabi, ruler of Babylon, to one of his governors in about 1800 BC.

waystations where horses were kept for the use of the couriers carrying oral instructions or messages inscribed on tablets.

The Roman *cursus publicus*

The Han Dynasty, which ruled China from the 2nd century BC to the 2nd century AD, had a sophisticated postal service, the very existence of which was unknown in the West. In contrast, much information has long been available on the Roman *cursus publicus*, set up by the Emperor Augustus in the 1st century AD.

The word *cursus* was used by the Romans to mean 'race', and it hints at the purpose of the network, which was to allow the rapid transmission of administrative dispatches across the empire. The couriers who carried them travelled at first on foot or on horseback, although in time chariots and carriages with iron-rimmed wheels came into general use, transporting not just messengers but also administrative functionaries provided with the necessary official passes. The post also served a military function, giving Roman legions the capacity to call for reinforcements and provide reports on the progress of engagements.

The system was made possible by the development of a Roman road network as remarkable for its efficient organisation as for its extent. Travellers were able to change mounts at regular stops along the way. Large privately owned mansions, operated by businessmen, provided food and lodging as well as stabling and the services of a blacksmith for the horses. Progress was generally between 40 and 65 miles a day.

Temporary decline

The advanced state of the Roman postal service stood as a monument to the administrative efficiency of the empire as a whole. The roads that the legionaries built often remain to this day as the main thoroughfares in lands that the Romans once ruled. As for the postal system itself, it survived Rome's fall at least until the 9th century, when the highways on which it operated began to fall into disrepair

and political fragmentation took away the need for a unified communications network. When the system finally came back into use from the 11th century on, it no longer served the purposes of a centralised administration. Local rulers still had to send messengers bearing administrative dispatches around their lands, but there were also now other, private individuals and institutions that needed to communicate across long distances. These included the great feudal lords, the newly founded universities and the monastic orders, which in many cases maintained houses in several different countries.

The post goes private

In the years before state postal services were opened up to private clients, the most developed courier systems were those run by the universities. Some operated

Breathless bronze A statue commemorating Pheidippides, the ancient Greek courier who inaugurated the concept of the marathon. In 490 BC, the legend goes, he ran to Athens with news of the Greek victory over the Persians at the Battle of Marathon. After delivering his triumphant message he collapsed and died from exhaustion.

Saintly rotula A medieval scroll containing a panegyric of St Vitalis, an abbot of Savigny in central France who died in 1122.

A LENGTHY READ

In the Middle Ages, communications between monasteries often took the form of scrolls, known as rotulae, which were passed on from abbey to abbey. At each stop on the way, the recipients would peruse the contents by unrolling the parchments, which could be as much as 20m long. They might then add news of their own before forwarding the scroll.

French address
An ornate 18th-century street sign from Chantilly, near Paris, bears the arms of the aristocratic Condé family (below).

AVX ARMES DE CONDE
POSTE ROYALLE

Heat treatment
Long before people began to worry about pandemics or bioterrorism, mail from plague-stricken regions was given heat treatment in ovens like this one to disinfect it.

teams of messengers whose job was to carry not just letters and documents, but also remittances from parents to support their children's studies. In return for cash payments, the bearers would also carry private mail. Merchant guilds similarly came to maintain private postal networks, linking, for example, the commercial centres of northern Italy with the cities of eastern France, where important annual fairs were held. The Fuggers of Augsburg, the Visconti in Milan and the civil authorities of the Republic of Venice all had mail services that kept them in touch with correspondents across Europe.

In the service of rulers

The best known and most extensive of the private medieval postal services was the Thurn and Taxis system, named for its family founders. This courier service between the city states had begun in 1290, but in 1498 Franz von Taxis was appointed postmaster to the Holy Roman Emperor Maximilian I with the right to carry both government and private mail throughout the Empire. From 1512, after being granted a monopoly by Maximilian, the

service opened up networks in Spain, Austria, Italy, Hungary and the Low Countries. It survived in private hands until 1867, when the service was nationalised by the Prussian state.

At the start of the 12th century the French monarchy depended on a single horseman and a few foot couriers to transport all its communications. By the reign of Charles V, 150 years later, that ratio had reversed with the court employing 36 mounted messengers and just eight on foot. Louis XI expanded the service further with the Edict of Luxies, which forbade postmasters, on pain of death, from handling any correspondence not expressly authorised by the king. The system became less rigid from the late 1400s on, principally for financial reasons: maintaining the relay stations proved expensive, and the postmasters were therefore instructed to bring in revenue by extending services to clients who could afford them. A century later Henry III created a royal messenger service that was open to private as well as public use, although still at a price. A delivery system specifically geared to carrying private citizens' mail finally saw the light of day in France in 1603.

The Royal Mail

In Britain the Royal Mail traces its origins to 1516, when Henry VIII created the position of Master of the Posts, forerunner of the later Postmaster General. Charles I opened the service to the public in 1635, and his son Charles II established the General Post Office following the Restoration of 1660.

The introduction of adhesive postage stamps in 1840 helped to accommodate the need for mass communication and also standardised the post. For the price of one penny a letter weighing ½oz (14g) could be sent anywhere in the country. Until sheets of stamps were perforated in 1847, they were cut up by hand by post clerks. The post in Britain has continued under royal patronage to this day, with trains and planes replacing the mail coaches that once criss-crossed the land. Today Royal Mail Holdings is a public limited company,

still dependent in part on finance from the state. It lost its monopoly on postal deliveries in the UK in 2006; its telecommunications arm split off in the privatisation of British Telecom 25 years earlier.

Dispatched by balloon
Sent from Paris during the siege of 1871, these letters successfully reached their destination.

Early stamps
The Penny Black of 1840 was replaced by the Penny Red in the following year.

THE FIRST POSTAGE STAMP

In the early days of private postal services the cost of postage was generally paid by the recipient. That changed when stamps were introduced, indicating that the sender had prepaid the cost of delivery. The first stamps were issued in Britain in 1840 following a proposal put forward by the social reformer Rowland Hill. Before that time as much as one-eighth of all mail had been dispatched under the personal frank of MPs and peers.

BY LAND, SEA AND AIR

In the early 19th century the post travelled in mail coaches – large, horse-drawn carriages that also took paying passengers and were considered faster and more dependable than ordinary stagecoaches. A first, experimental boat service came into operation between Dover and Calais in 1835. By that time rail delivery was also coming into use, the first dispatch having been made between Liverpool and Manchester as early as 1830. Mail trains took little time to kill off the mail coach; the last provincial coaches ceased to run in the 1850s. The main 20th-century innovation was the introduction of air mail; the first official delivery by aeroplane was made by a French aviator in India in 1911.

Rail mail *This pioneering mail van came into use in France in the 1840s.*

The contours of the known world

Just as writing preserves a society's memories, so maps serve to delineate the known world. The earliest known examples date back to the 3rd millennium BC, but it was only with the ancient Greeks that map-making took on the attributes of a science. Greek scholars can rightly claim the honour of laying the foundation of modern cartography.

The first steps
Drawn up in the 12th century BC, this Egyptian papyrus (top right) shows the location of gold mines in the Wadi Hammamat. The cuneiform tablet below it features a schematic representation of Mesopotamia, where it was produced in the 7th or 6th century BC.

Preserved today in the Italian city of Turin, an ancient Egyptian papyrus dating back to 1100 or even 1200 BC has a strangely familiar look. There is a patch of blue indicating the presence of water, mines outlined in red and black roads running from the River Nile into the surrounding desert. There are captions added by the scribe who drew up the manuscript to explain what the various elements of the drawing meant. Unmistakably, the document is a map, showing in miniature a landscape that in real life would have been too vast for the naked eye to take in.

Mastering space
The Turin map was created to fulfil a practical purpose. It was intended for use by an overseer supervising the transport of blocks of stone from mines. Throughout map-making's long history, one of its chief goals has been to put travellers on the right road, and the art has progressed in sophistication as the boundaries of the known world have been extended. By outlining the globe, maps have helped people not just to grasp but also to conquer its distances. No surprise, then, that maps go back to the birth of writing or perhaps even earlier, answering as they do a basic human need. As early as the 3rd millennium BC, Babylonian scribes were fashioning schematic representations of Mesopotamia on clay tablets. Even at that date frontiers were drawn in, suggesting that maps already served a political function.

In the 1st millennium BC Chinese officials of the Western Zhou Dynasty illustrated reports of their travels to remote parts of the empire with maps or topographical reliefs of the territory. Sadly, none have survived; we know of them from texts which provide clear evidence of the early development of cartography in China, where the science of map-making seems to have been closely linked to the management of canals and waterways.

A vision of the world
Like literature or the visual arts, maps say a great deal about the way in which their makers view the world. The oldest surviving Chinese geographical work is the *Tribute of Yu*, written in the 5th century BC. This put the nation's capital firmly at the centre of things. Beyond it lay first the imperial domains, then the vassal lands and a buffer zone of pacified tribes, which in turn gave way to the lands of

THE BIRTH OF GEOGRAPHY

Scholars have long considered Herodotus to be the 'Father of History', yet the famous Greek traveller, born in about 484 BC in Asia Minor, also has a claim to be the first true geographer. In the decade after 450 BC he set out to investigate the wars that the Greek city-states had lately fought against the Persian Empire. In the course of his researches, he travelled more than 3,000km across the Near East and into Egypt. Herodotus eventually synthesised the information he had gained in his *Histories*, which were not just as a record of events but also described the physical framework in which they took place. The geographical mindset that he pioneered enabled the Greeks to form the most coherent view of the world attained to that date.

World map
A map drawn up in ancient Greece shows the Eurasian landmass and North Africa surrounded by a band of ocean.

Mosaic map
Made in the 6th century AD, this mosaic shows the town of Jericho and the River Jordan, embellished with stylised palm trees.

barbarian peoples considered to be allies, and finally to regions inhabited by savages. The Chinese would continue to hold the same Sinocentric viewpoint into modern times.

The *Tribute of Yu* presented the world as a series of concentric rectangles, perhaps contributing to the subsequent development of a grid system in later maps. In China, the scale and orientation of maps were standardised in the 3rd century AD by Pei Xiu, a celebrated geographer who served as minister of public works under the Jin Dynasty. Thereafter,

thanks to his efforts, maps could be read and understood by people other than those who originally drew them up.

The first Greek maps

As navigators, explorers and conquerors, the Greeks gathered much geographical knowledge at first hand. They soon felt a need to put the information on paper. They also speculated about the possible shape of the Earth. Such questions first became serious objects of study at Miletus in Asia Minor in about 600 BC.

The first Greek known to have produced a map was Anaximander, a disciple of Thales, father of the natural sciences. Anaximander's map showed the world in the form of a flat disc with Greece at its centre. A circular band of ocean surrounded the central Eurasian landmass. The next known map, produced by Hecataeus a century later, followed in very much the same tradition.

The great traveller Herodotus, who lived in the 5th century BC, was highly critical of these early efforts, castigating them in particular for making Europe and Asia the same size. He also rejected the notion of the ocean as a vast stream encircling the world. In subsequent years Carthaginian ships probed up the Atlantic coasts of Europe as far as Brittany,

A schematic worldview

In the 12th century European map-makers were still reproducing centuries-old schematic views of the world represented in the form of a T (top), with Asia at the top, Europe in the bottom left-hand segment and Africa to the right. The map by Arab geographer Al-Idrisi (above), dating from the same era, gives a more realistic picture of the Mediterranean Sea with Egypt below it; the River Nile is marked in red.

adding greatly to geographical knowledge, as did information gained in the Persian Wars and Alexander the Great's campaigns in Asia.

Pytheas, Eratosthenes and Hipparchus

The most significant of all the voyages of exploration was that of Pytheas of Massalia (modern Marseille). In about 325 BC this Greek navigator sailed through the Straits of Gibraltar and circumnavigated the British Isles. Noting that the Sun hangs lower in the sky in northern latitudes, he took precise measurements to document the phenomenon.

In subsequent years Alexandria, the capital of Ptolemaic Egypt, became the centre of geographical learning. It was there that Eratosthenes (c276–194 BC) used Pytheas's observations to establish that the Earth is a sphere. Although the idea had previously been floated by Pythagoras and others, Eratosthenes was the first person to base his argument on firm astronomical evidence.

Even so, his work had limitations. In 139 BC Hipparchus, the greatest of all the Greek geographers, criticised Eratosthenes for basing his calculations on inadequate data. Hipparchus went on to propose the model of the Earth used to this day, seen as a sphere divided into 360 degrees of longitude. At the time this approach caused difficulties, for while latitude could be calculated with reference to the angle of the Sun, longitude could only be worked out with the aid of accurate chronometers, which were yet to be invented. Hipparchus himself suggested using eclipses, a method that lacked precision.

Ptolemy's successes and failures

Cartography remained an inexact science when, in about AD 150, the geographer Ptolemy sought to correct the situation by producing a work that brought together all the cartographical data then available. Yet despite Ptolemy's genius, his work proved fallible.

The problem lay not so much in Ptolemy's method as in the inaccuracy of the observations he relied on, which came from travellers and military men who had recorded them without much precision. He overestimated the extent of Eurasia, projecting it over 180 degrees of the Earth's circumference instead of 130; he also set the Equator too far north and got almost all of his lines of latitude wrong, except for the island of Rhodes. Through no fault of his own, lack of knowledge meant that Ptolemy was totally unaware of the existence of Australia or the Americas. Most significantly, he turned his back on Eratosthenes's calculations of the

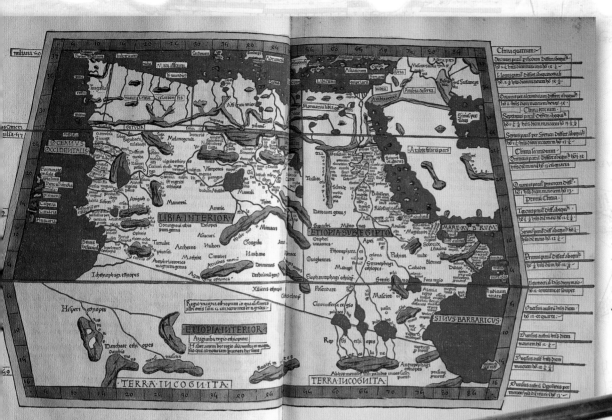

Ptolemaic view
A map created in the 15th century showing North Africa and the Arabian Peninsula still largely reflects the worldview of the Classical Greek geographer Ptolemy.

circumference of the Earth, preferring to go with those of Posidonius of Rhodes, which had shrunk the globe by at least a third to a circumference of just 28,000km.

A rich heritage

But sometimes errors can unexpectedly produce results. Some 13 centuries after Ptolemy made his mistakes, they contributed to the discovery of America by Europeans, encouraging Christopher Columbus to underestimate the actual distance from the Atlantic coast of Europe to the East Indies, the explorer's intended destination.

Columbus had access to Ptolemy's work thanks to the Arabs, who preserved it through the European Dark Ages and added to the stock of knowledge through discoveries made by their own merchants, travellers and conquering armies. In the 12th century – when Christian Europe's worldview was still largely theological based on ideas outlined 500 years earlier by St Isidore of Seville – King Roger of Sicily was able to turn to the Morocco-born Muslim geographer Muhammad al-Idrisi for a comparatively realistic map of the Mediterranean Basin. Al-Idrisi drew heavily on Ptolemy's work, which continued to serve as a reference point for scholars until it was finally outdated by the new knowledge born of the great voyages of exploration during the 15th and 16th centuries.

Royal astrolabe
Astronomical or navigational instruments like the astrolabe played a crucial part in improving geographical knowledge. This one (above) was once the property of Peter the Great, tsar of Russia in the 18th century.

THE FIRST RELIEF MAPS

Military men took a professional interest in maps from the earliest times. They were particularly concerned with mountains and rivers, which could aid or block an army's progress. Among the best-known of all early maps was the one prepared in 210 BC for the tomb of the First Emperor of China, Qin Shihuangdi, which had hidden mechanisms that made the rivers flow with mercury. Relief maps became relatively common in China; one prepared in AD 32 for a general named Ma Yuan featured mountains modelled out of rice flour. In the 5th century a map of all the imperial lands was fashioned out of wood, showing mountains and rivers along with demarcations of the various regions.

Marine highways for commerce and communication

Sea trade was already up and running in the Mediterranean and the Red Sea by the start of the Iron Age. In the centuries that followed new routes were pioneered in the Atlantic and Indian oceans, as fresh vistas opened up through commercial contacts, imperial expansion and the discoveries of intrepid explorers.

Greek bireme
Two banks of rowers powered this ancient Greek vessel, the bottom row using portholes in the hull. The sail was lowered for combat.

In the 12th century BC the Phoenicians in their pot-bellied ships ruled the waves, at least in Mediterranean waters. Having started out as proxy partners carrying freight for Egyptian and Assyrian merchants, they began to operate trading posts of their own from the 10th century BC, when the port of Tyre first rose to prominence. Before long they had bases on Cyprus and other Mediterranean islands as well as in southern Spain, a rich source of silver. The Phoenicians also established ports of call all along the North African coast, from Egypt to the Straits of Gibraltar, each about 30km from its neighbour – a day's sailing, to avoid having to navigate by night. In 814 BC a new maritime power, Carthage, was founded on what is today the Tunisian coast; by the 7th century BC it had come to control the southwestern Mediterranean region.

Phoenician trireme
In the 8th century BC triremes featured three banks of rowers (below).

PIRATES AHOY!

Envious eyes coveted the merchandise carried along the sea lanes, adding piracy to the long list of dangers facing those at sea. In the Mediterranean region the inhabitants of Aetolia on the Corinthian Gulf, Sardis and Cilicia in Asia Minor and the islands of Crete and Corsica were all notorious for their raiding. Respectable citizens in Rome and Phoenicia were even happy to employ pirates to do their bidding. From time to time action was taken when the scourge of piracy threatened order and prosperity. Thus in the 5th century BC, Athens dispatched its war fleet against the pirates of the Aegean, and in 67 BC the Roman general Pompey waged war on Cilician sea raiders at the head of a force of 5,000 galleys and 120,000 men.

The Greek Mediterranean

By that time a rival power had emerged in the shape of Greece, which was recovering from its Dark Age and once more taking to the sea. Traders and colonists alike set out for the western Mediterranean and east to the Black Sea in bigger-sailed ships that were more streamlined and handled better than their Bronze Age predecessors.

Technological progress emboldened sailors, who by the 7th century BC were happy to spend nights at sea, their boats securely moored by double-pronged 'fisherman's' anchors. They gained extra confidence from the 2nd century on, when new instruments became available enabling a course to be plotted against the heavens, so gaining a clearer picture of a ship's actual position at sea.

On the one hand mariners now had access to the first, rudimentary astrolabes, and on the other to reasonably accurate ephemerides – astronomical tables noting the changing position of the heavenly bodies month by month. The combined effect was to make sailors less nervous of venturing out of sight of land and more willing to risk night-time navigation. By that time, too, the first lighthouses had been built to help guide vessels safely into ports. Few were more frequented at the time than Piraeus, where shipments of corn from Italy and Scythia were unloaded to feed the neighbouring city of Athens.

In the 3rd century BC, in the wake of Alexander the Great's globe-trotting exploits, the principal Mediterranean sea-routes shifted toward Ptolemaic Egypt. The great port of Alexandria became a centre for the trans-shipment not just of Egyptian cotton and cereals, but also of African ivory and gold, as well as perfumes and spices from Arabia and India.

A Roman lake

The Mediterranean was never busier in antiquity than under the Roman Empire in the first five centuries of the Christian era. Rome's formidable war galleys quickly enabled it to establish maritime supremacy over its rivals; Carthage was destroyed in 146 BC and thereafter the victors took to referring to Mediterranean waters as *mare nostrum* – 'our sea'. The heavy Roman merchantmen, equipped with decks and two squared-off sails,

could make the journey from Ostia to Syria in 20 days, bringing back oriental silks, spices and iron. Others headed west for Gaul or Spain, both Roman provinces; it took only two or three days to reach Massalia (modern Marseille), which itself conducted a busy trade with ports on the North African coast, in particular with Alexandria. The sea-lanes were busiest in summer, when mariners took advantage of the favourable weather. In contrast, maritime traffic all but closed down between mid-November and March, when conditions were too dangerous for the big boats to venture far from shore.

Greek war galley
A model of a 9th-century-BC Greek warship (left). Sometimes called a monoreme, the galley had a ram that could penetrate an enemy vessel's hull.

Roman bridge
A wall painting unearthed in Pompeii (above) shows the bridge of Stabiae (modern Castellammare, near Naples) as it looked in the 1st century AD.

Arab traders

*A manuscript
illumination
(opposite page, top)
gives a stylised view
of the vessels that
criss-crossed the
Persian Gulf in
the 13th century
carrying cargoes
of ivory and spices
(opposite, bottom).*

A time of anarchy

The collapse of the Roman Empire following the barbarian onslaughts of the late 5th century AD left the western Mediterranean at the mercy of pirates and commercial traffic declined. In the east, the Byzantine Empire at first maintained order and prosperity from its capital of Constantinople (modern Istanbul), thereby benefiting the sea traders of the Italian city-states of Venice, Amalfi and later Genoa, all of which maintained close commercial relations with the Eastern lands.

In time, fresh sails appeared on the horizon, belonging to the lateen-rigged vessels of the Saracens. By the 9th and 10th centuries much of the Mediterranean was in Arab or Moorish hands, from Almería in Spain to the ports of the Levant.

The Indian Ocean

As early as the 10th century BC Egyptian mariners had reached the Land of Punt, thought to have been in the Horn of Africa. In so doing they opened up the Red Sea for trade, perhaps travelling as far as Mozambique in search of gold and spices. Another four centuries passed before the Persians and Phoenicians pioneered the sea route from the Red Sea to India. At first controlled by the successors of Emperor Darius, then by the Egyptian Ptolemies, Sassanid Persia and the Romans, it came to serve as a conduit for all the treasures of the Orient, from Malabar pepper and central Asian iron to Chinese silks.

The maritime kingdoms founded in the south of India from the 7th century on further boosted trade with Burma, Indochina and the Malay lands, as sailors learned by trial and error how to take advantage of the monsoon winds blowing from the northeast in winter and from the southwest in summer. In time, maritime traffic adapted to the rhythm of the monsoon, which favoured long journeys when conditions were right, but imposed lengthy stays in port waiting for winds to change. Return journeys to and from the Red Sea or the Persian Gulf could take more than a year.

From the 1st century AD Chinese junks, with their squared-off hulls and battened sails, ventured from their home waters on the long

THE PACIFIC VOYAGERS

The Pacific and its 10,000 islands remained beyond the main highways of marine commerce until the 19th century, but before that time the ocean had seen almost 9,000 years of practically uninterrupted sea migrations. The bulk of the traffic moved eastward from the Southeast Asian landmass. The first seafarers probably travelled by raft, although double-hulled canoes, sometimes equipped with a mast and sail, were used later. Travelling out from the mainland or from the Indonesian islands, the first emigrants became the progenitors of the peoples of Melanesia and Micronesia, as the two great island groups of the southwest Pacific are now known. From about 1700 BC a fresh wave of exploration carried people from Melanesia and elsewhere into unknown waters further east, where they settled the islands of Polynesia, reaching the last of them, Hawaii, perhaps as late as AD 750.

Long-lasting style
A Chinese junk in Hong Kong harbour today looks little changed from its predecessors of 500 years ago.

Conquest carried the Arabs from their original desert homeland to coastal regions where they came into contact with maritime peoples. They soon developed an appetite for sea travel, borrowing their classic vessel, the dhow, from the inhabitants of Ethiopia's Red Sea coast. Heading south, they set up trading posts on Africa's east coast as far south as Sofala in what is now Mozambique.

Trade in the Indian Ocean received a further boost in the years after AD 762, when Baghdad was founded to serve as the Caliphate's new capital. Muslim sailors and merchants, both Arab and Persian, sailed from Siraf to the Laccadive and Maldive islands, then on to India, Sri Lanka and southern China, which they first reached in the 8th century. On the outward journeys the boats carried ivory, incense, copper, rhinoceros horn and slaves; they returned loaded with all the riches of the Far East.

In the cosmopolitan port of Canton (modern Guangzhou) the visitors had leisure to admire the great multi-decked, sea-going junks that the Chinese had been constructing since the 10th century. They took from their hosts the idea of the central, stern-mounted rudder, which arrived in the West in the 12th century along with the magnetic compass, another Chinese invention.

journey to India. There, in the ports of the Coromandel Coast, they unloaded Buddhist pilgrims along with silk, packed for the voyage in compartmentalised holds, before returning with cargoes of amber, glass, agate and carnelian. The vessels were equipped with stern-post rudders, and by the 3rd century they also had obliquely rigged fore and aft masts, which enabled them to sail into as well as before the wind. These advances allowed Chinese mariners to set their sights on more distant horizons. From the South China Sea they opened a regular sea-lane to Sumatra in the East Indies. Closer to home, by the 7th century they were making voyages to Taiwan and the Ryukyu Islands south of Japan. Two hundred years later they could be found in the Indian Ocean, travelling as far as the port of Siraf in the Persian Gulf, where they came into contact with Arab shipping.

Dragon boat
A model of the Gokstad boat, excavated in Norway in the 19th century, shows the swelling lines and elegant construction of the Viking longships (right).

Ra II
Thor Heyerdahl's reed boat in rough seas off the Moroccan coast (below).

The Atlantic before the Vikings

According to Herodotus, Phoenician sailors in the service of the Egyptian Pharaoh Necho II sailed in about 600 BC from the Red Sea around the coast of Africa. If his account is correct, their trip would have been the first Atlantic voyage known to history. There is more certainty about the exploits of Hanno, a Carthaginian explorer. In about 465 BC he led a fleet of 60 ships to establish colonies on Morocco's Atlantic coast, subsequently sailing as far as Madeira and the Canary Islands.

At roughly the same time, Hanno's compatriot Himilco ventured north into the seas around Britain. This is the route that Pytheas, a Greek from the colony of Massalia (Marseille) retraced in about 325 BC (see page 48).

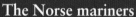

Pytheas extended it to the coasts of a land he called Thule, now thought to have been either Iceland or northern Norway. Once the northern sea lanes had been opened, sailors from Phoenicia, Carthage and Greece established trading links with the British Isles, a rich source of tin, as well as with lands bordering the North Sea and the Baltic, where amber came from.

The Norse mariners

The Mediterranean powers dominated these routes until the late 8th century AD, when the Viking dragon boats first made their presence felt in Atlantic waters. To explore new routes the Norsemen generally preferred the smaller, more manageable knarrs, merchantmen about 20m long. Lacking maps, the pioneers mostly stayed close to the coast, looking out for familiar landmarks. When out of sight of land, they relied on close observation of natural phenomena. By such means they settled the Faroes by 700 and reached Iceland in 815. One settler there, Erik the Red, went on to discover Greenland in 982. Some years later his son Leif Eriksson ventured further west in search of a land reported decades earlier by the crew of a ship swept off course by a storm. Leif duly reached North America, landing in Labrador and then Newfoundland. No colony was established and the memory of his discoveries was largely forgotten. The trans-Atlantic sea route would remain unopened for centuries to come.

THOR HEYERDAHL, THEORIST-ADVENTURER

In 1947 the Norwegian ethnographer Thor Heyerdahl sailed 6,500km from Peru to Polynesia on the *Kon-Tiki*, a balsawood raft constructed using traditional Peruvian boatbuilding methods. His goal was to prove that early mainland peoples had the technology necessary to cross the Pacific Ocean, raising the possibility that they might have made contact with the islands in ancient times. Twenty-five years later he sailed *Ra II*, a 15m-long boat made of papyrus reeds, from the Atlantic coast of Morocco to Barbados. The trip not only demonstrated the seaworthiness of such vessels but also showed that links could have existed between Africa and South America long before Columbus's day, perhaps explaining perceived similarities between ancient Egypt and ancient cultures of Mesoamerica. The theory remains controversial, but Heyerdahl's achievement is still feted.

Team effort
Chinese women of the Song Dynasty, at the court of the 12th-century Emperor Huizong, use a device shaped like a bed-warming pan to smooth wrinkles from a length of cloth.

The iron c400 BC

Long before the 4th century BC, when the Chinese invented the iron, people were improvising presses from flat stones or blocks of wood to smooth out animal skins. Similar implements, fashioned in later times from marble or glass, continued to be used up to the 15th century AD to remove creases from linen, since the employment of gum to stiffen adornments like ruffs meant that delicate garments could not withstand heat treatment.

The first Chinese irons resembled brass warming-pans that were filled with burning embers, charcoal or hot sand. To judge from the finely pleated linen depicted in ancient Egyptian art and statuary, they too must have had some sort of implement for pressing clothes, at least from New Kingdom times in the 16th century BC. There is no clear evidence to show when irons first reached the West, but by the 15th century AD Dutch tailors were using steel devices containing a compartment for a piece of fire-heated red-hot metal.

From the 16th century a choice of irons of different types became available. Flatirons – thick, triangular-shaped implements with a handle, that could be heated on the hearth – remained in common use until the 19th and even the 20th century. From the 17th century people also used box irons that took the form of an openwork iron container, with a wooden handle, that was filled with hot embers. Similar contraptions were made in terracotta, provided with ventilation openings to keep the ashes glowing.

ELECTRIC IRONS

The classic 19th-century laundry iron took advantage of another innovation of the day – the cast-iron stove, which provided a ready source of heat within the home. A new era dawned in 1882 when the American inventor Henry W Seely introduced the electric iron, even though few people could use one at the time for want of a household electricity supply. For the next few decades Seely's new device faced competition from irons heated by liquid fuels including natural gas, alcohol, kerosene and whale oil; these implements were sold in rural areas of the USA right up until the Second World War. The electrically powered steam iron, which produced improved results by moistening the washing, was an American invention of the 1920s that was soon itself upgraded by the addition of an adjustable thermostat.

Hot iron
Two young Pakistani girls blow on the embers used to heat an old-fashioned box iron still in use today.

The pulley
c400 BC

Jacob's well
A 6th-century mosaic depicting Jesus conversing with a Samaritan woman at a well shows an early pulley in operation.

marine use were among the first objects to be mass-produced in the Portsmouth dockyards, from 1803 on, employing machinery designed by Marc Isambard Brunel, father of the great Victorian engineer Isambard Kingdom Brunel.

Pulleys took on a new importance with the coming of the Industrial Revolution. Previously used principally for lifting, now they were employed as drive belts. Power from a centralised steam engine was transmitted to the factory machinery via thousands of pulleys, at first made of cast iron and later of steel. In recent times, with the development of synthetic materials, pulleys have once more started to replace gears and chains, particularly in the automobile industry.

Tradition ascribes the invention of the pulley to Archytas, a philosopher and mathematician who lived in the Greek colony of Tarentum from 438 to 365 BC. Yet the Erechtheum was built on the Acropolis at Athens when Archytas was a youth, and some say pulleys were used in its construction. The first written description of the device occurs later, around 330 BC, in the *Mechanica*, a work long attributed to Aristotle. The inspiration came from the wheel, but perhaps also from the fixed pulley employed earlier by the Egyptians, which consisted of a wooden beam or block of stone notched with a groove shaped to take a greased rope.

Archimedes (287–212 BC) is said to have invented the block-and-tackle pulley, designed to provide added leverage and so reduce the amount of effort required to lift objects. As technical adviser to Hiero II, king of Syracuse, Archimedes was presented with a seemingly insoluble problem. A ship named the *Syrakosia*, the biggest of its day, became stuck on the ramp at its launch. Given this chance to test his theories, Archimedes devised a system of pulleys and levers that successfully shifted the vessel. According to legend, his solution was so effective that Hiero himself was able to activate the mechanism single-handed.

Lifting and carrying

By Renaissance times pulleys were principally used on board ships; as many as 1,000 were fitted on large galleons. Wooden pulleys for

Public works
The wooden pulley (above) was one of many used in the construction of the Duomo in Florence, the famous dome-topped cathedral, in the 15th century.

Lifting gear
A pulley eases the task of drawing water from a well in Tunisia (right).

Gears
c300 BC

The mechanism of gears involves at least two toothed wheels that mesh with one another to transmit motion from one to the other. The wheels are usually of different sizes and so move at different speeds, exerting greater or lesser torque.

Like the pulley, this invention dates back to ancient Greece. Gears were mentioned in the *Mechanica*, a 4th-century-BC work long attributed to Aristotle but now usually considered the work of another, unknown author. The first known applications of gears date from the 3rd century, in the clepsydra (water clock) designed by Ctesibius of Alexandria and in Archimedes' screw.

The teeth of progress

From the start, gears came in two main types, as they do to this day. On the one hand were large-scale mechanisms, generally made out of wood, that were essential to the functioning of lifting machines and weapons of war, and later for mills, vehicles and industrial machinery. On the other were the small metal cogs, beloved in precision engineering, that over the years have been employed in clocks, automata and calculating machines.

Gearing can be used to change the direction of a driving force. Before the Industrial Revolution, one common arrangement had the teeth of a gear wheel meshing with the spaces between small parallel bars joining twin wooden discs, the whole forming a hollow cage. In water mills, gears transferred the vertical movement of water falling onto the vanes of the wheel into the horizontal turning of a millstone. The sakias (water-raising devices) of Egypt worked in a similar way, raising water with the aid of a vertical belt powered by the horizontal motion of a draft animal.

CHAIN TRANSMISSION

Philo of Byzantium was the first person to describe chain transmission, in the context of a chain drive connected to a windlass used to fire arrows from a repeating crossbow, in the 2nd century BC. A similar mechanism next appeared in China in AD 976, used by Zhang Sixun in the escapement mechanism of his astronomical clock. French inventor Jacques de Vaucanson employed chain transmission in the 18th century to help power the world's first automated loom. It became part of everyday life in the late 19th-century Europe, when bicycles equipped with gears and a chain replaced earlier boneshakers.

Slow gear
First invented by Greek engineers and later, as the 13th-century manuscript (above) suggests, adopted by the Arabs, gears were slow to be generally adopted. They proved their worth above all in mills, being used to transfer and direct wind or water power through arrangements like this one (top) in a windmill in the French province of Anjou.

The sailors' friend

As a maritime signalling system giving warning to sailors, lighthouses have been known since antiquity. They have evolved from fiery beacons lit on promontories to today's completely automated machines. The most famous of all was the Pharos of Alexandria, one of the Seven Wonders of the Ancient World.

Battered by waves
The Kéréon light stands far out to sea southeast of the island of Ushant off the Brittany coast (main image).

With only the stars to guide them by night, the mariners of antiquity feared overcast skies as much as they dreaded storms. Consequently, as early as the 1st millennium BC, the Phoenicians developed the practice of lighting fires at strategic points along their coastline to serve as beacons for navigators.

The first lighthouses

Both the Greeks and Romans built towers designed to perform the same function. One of the earliest known to history lit up the promontory of Sigeum on the Asian shore of the Dardanelles Strait from the 9th century BC. By far the best-known of these ancient lighthouses was constructed on the island of Pharos at the entrance to the harbour of Alexandria, to designs by an architect named Sostratos of Cnidos. This extraordinary white-marble structure, built in about 285 BC, stood some 130m tall and was topped by a statue of the sea god Poseidon. A perpetual

Pharos of Alexandria
The great lighthouse of the ancient world is depicted on this 2nd-century-BC Roman coin (above).

Lighthouse relief
A Roman bas-relief of the 3rd century BC (below) shows the lighthouse at Ostia, the port of Rome, framed between two merchant vessels packed with amphoras.

flame burned at its apex, the light from which was amplified by a reflecting mirror.

By the 4th century AD, when their empire was approaching its end, the Romans had built some 30 lighthouses from the Black Sea to the shores of the Atlantic. The amount of traffic using the sea lanes was much reduced after the fall of Rome, and the lighthouse network fell into disrepair. No new lights were built, and even the great Pharos of Alexandria gradually deteriorated; it was finally brought down by an earthquake in the 14th century.

Shedding more light

In ancient times beacon-tenders burned resinous wood that gave off clouds of smoke, which increased the distance from which the warning fires could be spotted. By the 1st century AD their successors were using oil lamps. In 1782 a Swiss inventor named Aimé Argand introduced improvements that increased the amount of light given off by oil lamps by at least a factor of ten; the secret lay in a cylindrical wick that was topped by a metal chimney to enhance the air flow.

THE CORDOUAN LIGHT

One of the oldest lighthouses still in active use is sited on the tiny island of Cordouan off the mouth of the Gironde estuary in southwestern France. The history of the Cordouan Light can be traced back to the Middle Ages, when Moors from Córdoba in Spain established a trading post there, lighting fires as signals for their shipping. By the mid 14th century a polygonal tower had been built, with a beacon on top that was tended by a hermit. Constantly battered by wind and waves, the building fell into disrepair and was rebuilt in the late 16th century. In 1823 Cordouan became the first lighthouse to use a Fresnel lens. Since 1896 it has housed a rotating drum containing eight prismatic lenses backed by mirrors, giving out a light that is visible more than 50km away – over three times the previous distance.

Over the following century lighthouses were provided first with gas and petrol burners and then with arc lights before finally settling for electric bulbs. At the same time decisive progress was made in the field of optics. The French physicist Augustin Fresnel (1788–1827) made a crucial breakthrough when his duties as a commissioner for lighthouses suggested a practical application for theories he had already published on the subject. Fresnel realised that the lens used to project the beam actually absorbed much of the light energy produced. To remedy this he came up with a device, still known as the Fresnel lens, which breaks up the convex refractor into a series of recessed concentric rings. These concentrate the beam, thereby increasing threefold the distance from which it can be seen.

Light concentrator
The Fresnel lens (below) was first put to use in the Cordouan lighthouse (right). The lens below is in use in the Skellig lighthouse off the southwest coast of Ireland.

The path to automation

For most of their history, lighthouses have only functioned thanks to keepers who operated the light, thereby helping to assure the safety of those at sea. In the second half of the 20th century, this ancient profession virtually disappeared as lighthouses were automated. Often employing wind or solar power backed up by an emergency generator, lighthouses today are packed with complex electronic equipment allowing them to perform their important function unmanned.

WARNING LIGHTS AND FOGHORNS

Sailors recognise different lighthouses by the beams they emit. Each one has its own signature, made up of the colour of the light, whether it is fixed or flashing, and, in the latter case, by the interval between flashes. These visual signals are often supplemented by warning klaxons, mostly used in foggy conditions.

The bit, saddle and stirrups
c15th and 5th centuries BC and 3rd AD

The consensus of scholarly opinion holds that the horse was probably domesticated by about 3000 BC, and from that time on riders naturally sought to make the most of the animal's exceptional stamina, strength and speed. That meant learning to control the horse. The first pieces of equipment devised to that end were the bit and bridle, indispensable aids for communication between rider and horse. Starting off as a simple cord passed round the horse's lower jaw, the bridle was soon made more effective by the addition of the bit, fitted in the horse's mouth. The oldest known bits have been found in Scythian graves; carved at first from antler or bone, they were made of metal from about 1400 BC on.

Saddling up

The saddle was in comparison a late arrival. Horsemen for many centuries either rode bareback or else seated on pads of fabric or animal hides held in place by girths. The earliest-known proper saddles were found at burial sites in the Altai Mountains on the

Full riding tackle
A 12th-century Chinese nobleman enjoys the benefit of a saddle, stirrups, bridle and bit.

THE FIRST MOUNTED WARRIORS

Stirrups reached western Europe in about the 8th century by way of the Byzantines, who had themselves learned of their use from the Avars who had settled in Hungary two centuries before. In combination with the saddle, stirrups revolutionised the art of war, giving cavalry a predominant role that would not be challenged for centuries to come. The development had important consequences for medieval society, encouraging the emergence of an aristocratic class who had mastered the skills of horsemanship and could afford the considerable expense of running stables.

Russo–Chinese border. Used by horsemen of the Central Asian steppes from the 5th century BC, they consisted of two padded leather cushions mounted on a wooden frame. Knowledge of the invention spread slowly and selectively; by the following century the Persians were aware of it but the Greeks, their neighbours, were not. Like the bit, the saddle probably reached western Europe in the 4th century AD, brought by the Huns.

A safer ride

Stirrups were a Chinese invention of the 3rd century AD, at least in their modern form. As early as the 2nd century BC, barefoot riders in India had used rings just big enough to hold the big toe. By providing a stirrup with room for the whole foot, the Chinese greatly improved security and stability for the rider. By medieval times, knights riding into battle depended on their stirrups to lessen the risk of falling off.

Fancy bit *Made of iron and decorated with coloured glass, this Celtic horse's bit was made in the 3rd century BC.*

From the noria and sakia to Archimedes' screw

Ever since the beginnings of farming, the need to irrigate fields of crops in dry regions has created a demand for machines that raise water. The Archimedes' screw and the subsequent development of the noria lightened the burden for farmers and opened up new technological horizons.

Demonstration piece *A small ball placed at the bottom of this coil (above) climbs up the spiral, against gravity, when the handle at the top is turned. The model was made in the 18th century specifically to explain how the Archimedes' screw works.*

Sucking up water *Invented more than two millennia ago, screwpumps are still used in modern times for irrigation. This one (left) is in the Nile Delta.*

The year was 200 BC, and the late-afternoon sun was still burning hot. In the Greek-speaking lands of the eastern Mediterranean the ground was bone-dry and plants were wilting for want of rain that the heavens refused to supply. The irrigation canals and ditches that criss-crossed the fields contained at best a trickle of water.

A peasant crossed his arid field, making his way to the bank of a nearby river. There, he gripped the handle of a machine that had been recently installed and started to turn it. Before long, after just a few turns, a life-giving jet of precious water spurted into the artificial irrigation channel dug out at his feet – and continued to do so as long as he turned the handle.

Previously that same farmer would have raised the water with the aid of a shaduf – an upright frame supporting a counterweighted pole used as a lever to pull up water in a bucket. Although easier than hauling up a bucket on a rope, the effort involved in using a shaduf was still considerable, and each bucket of water raised had to be carried to the irrigation canal. The new device made life much easier as the whole job could be done simply by turning a handle: each revolution sucked up river water into the long tube. To the peasant, the invention must have seemed miraculous in its simplicity. A scholar might have told him that the mechanism that drew the water up was a screwpump.

Not Archimedes after all

The fact that we know this device by the name of the Greek engineer and mathematician Archimedes stems from a simple error traceable to a false commentary dating from the 17th century. Archimedes may have perfected the screwpump, but it owed its

ARCHIMEDES' GIANT PULLEY

It is no offence to Archimedes' memory to point out that the great Greek savant, who lived from 287 to 212 BC, did more to improve existing mechanisms than to invent new ones. In reaction against the philosopher Aristotle, Archimedes rejected abstract theorising and put his faith instead in observation and experimentation. Tradition ascribes various extravagant inventions to his credit. Plutarch claimed that Archimedes devised a gigantic pulley equipped with an equally large hook, which was fitted on warships for the purpose of seizing enemy vessels and lifting them bodily from the water. Such a device would in fact have required an enormous counterweight – one heavier than any that could have been handled by the naval technology of the day.

genesis to an earlier inventor, Archytas, who lived in the Greek colony of Tarentum in the 4th century BC.

The principle involved is straightforward enough: a revolving spiral will draw up air and liquids, such as water. To work in the field, the screw is enclosed within a hollow cylinder inclined on a river bank, ideally at an angle of 37 degrees, with the lower end immersed in the water. By turning a handle attached to the central axis, the operator sets the spiral blade in motion, and this draw water up to spurt out through an opening at the top.

Other lands, other uses

The invention of the screw not only simplified the task of filling irrigation ditches, but also allowed users to bring water up from deeper sources than had previously been possible. It provided some protection against drought as it allowed reservoirs to be constructed in the fields as well as outside cities. And it was not long before other applications were found for Archimedes' screw. The Romans used it to pump water out of mineshafts, where it proved

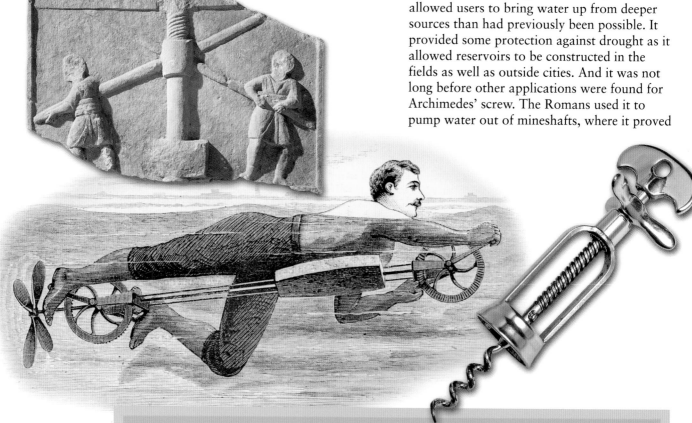

Harnessing a principle
The screw principle has been applied in many different contexts, from oil presses (top) to the bizarre 19th-century swimming machine (above) to corkscrews. No doubt there will be new applications dreamed up in years to come.

PRESSING, DRILLING, SHIFTING

Very soon after inventing the mechanical screw, the Greeks were finding fresh uses for the device and putting it to work in olive and wine-presses. In Renaissance times the technology showed up again in early printing presses. In each case the screw mechanism multiplied the amount of energy involved in turning it, causing a wood or metal block to press down with increased force on the objects placed beneath it – whether olives, grapes or, in the case of the printer's press, a sheet of paper brought into contact with previously inked characters. Miners' drills, cabinet-makers' augers and some corkscrews employ the same principle, using circular motion to pierce material with minimum effort, then drawing debris back up the spiral, just as the early Archimedes' screws drew up water.

When a combine harvester cuts crops and separates the grain from the chaff, an endless screw draws the grain to the grain tank. Similar arrangements are used to store grain in silos. Millers put the principle to work to shift grain and, later in the milling process, flour. Screwpumps can in fact be used to move any particulate matter that flows in the same manner as liquids.

more effective than the suction pump, introduced by Ctesibius of Alexandria at about the same time. The pump worked only intermittently with the rise and fall of the handle, but the screw's action was continuous.

The noria and sakia

Since the start of the 2nd millennium BC, long before Archimedes' screw was invented, people in the Near East lifted water with the help of a noria – a large wheel with buckets fixed to the rim. Initially they relied on brute strength to turn the wheel by pushing on a wooden beam. By the time the screw came into use, norias were self-propelled, at least in Egypt, thanks to wooden paddles attached to the wheel rim which harnessed the force of the current to make the wheel turn.

In the 2nd century BC a more complex form of noria made its appearance, this one powered by a draft animal. The animal trudged in a circle, hitched to a horizontal driving wheel that was geared to make the bucket-bearing vertical wheel revolve in its turn. In Egypt, where the device was probably first developed (some authorities prefer the claim of Syria), it was known as a sakia.

Sakias were more powerful than the earlier norias, and could be used to draw water from watercourses with only a feeble current or even from pools with no flow at all. Such devices remained in use for millennia; a century ago they could still be found across the Near East as far as the Indian subcontinent. They may well have provided the inspiration for the watermills developed by the Romans.

Developments and improvements

Paddle-powered norias and sakias turned by draft animals were only suitable for tapping ground-level supplies of water. To draw water from underground wells peasants still had to haul up buckets on the end of a rope or use some sort of shaduf arrangement – unless they lived in Greece or in China, where an improved sakia in the form of a chain pump made its appearance in the 3rd century BC. An endless chain fitted with discs to trap the water carried the liquid up a tube to ground level.

Used principally for irrigation, the various types of noria spread into Africa and around the Mediterranean lands in the 7th and 8th centuries AD, in the wake of the Arab conquests. They also provided water for city-dwellers, as at Hama in Syria, which was supplied by an aqueduct fed by waterwheels driven by the River Orontes. To this day norias are still in use in Egypt and the Middle East.

Endless circle
A draft animal powers a sakia, a form of noria still used to raise water in Egypt and other North African countries. The animal walks round in a circle turning the horizontal wheel, and the energy from this is conveyed by gears to the vertical wheel raising buckets of water.

ARCHIMEDES – *c*287 TO 212 BC
Mathematician of genius and practical engineer

Archimedes of Syracuse remains the best-known of all the ancient Greek scientists. He was one of the first people to marry mathematics with technology and bequeathed an outstanding legacy of practical discoveries, as well as the founding texts of physics.

The Eureka experiment

Artists through the centuries have imagined the scene as Archimedes tested the density of solid bodies by plunging crowns of equal weights but different materials into a container of water. They have spawned some misunderstandings along the way. Few illustrators realised, for example, that the silver crown would have been bigger than the golden one, said to have taken the form of a laurel wreath (right).

When Archimedes stepped into the bath, he was no doubt relishing the prospect of a few minutes of relaxation. The great scientist had spent the day pondering a delicate problem presented to him by Hiero, ruler of Syracuse. Still preoccupied as he lowered first one and then the other leg into the welcoming water, he was probably barely aware of splashes slopping over the rim. But as he sank deeper it dawned on him that his limbs seemed to be growing lighter as the water rose in the tub. A flash of inspiration then lit up his brain.

The Eureka moment

By now Archimedes' mind was buzzing. He realised that the solution to the question puzzling him lay in the water that was forced from the tub by the weight of his body. *Eureka*! He had the answer to the conundrum. Now he had to confirm as quickly as possible the truth of the idea that had just come to him. He leaped and ran, dripping wet, out of the bathhouse and into the street – stark naked, if we can believe Vitruvius, who tells the story in the ninth book of his *De Architectura* ('On Architecture').

Vitruvius may have enhanced the story in the telling, but Archimedes' discovery was certainly a dramatic one. Hiero had given the Greek scientist the task of unmasking what he suspected might be a daring fraud. A goldsmith had made the king a magnificent golden crown, but according to some malicious tongues he had held back some of the gold that the ruler had provided for the crown and replaced it with a similar weight of silver. Archimedes was charged with finding out if indeed such a deceit had been committed – but without in any way damaging the crown, which was dedicated to the gods of a temple in Hiero's capital at Syracuse.

Archimedes' bathtime inspiration was the realisation that, just as the volume of his own body had forced an equal volume of water to overflow from the tub, so any other object plunged into a container full of water would do the same. By measuring the crown's volume compared to its weight, he could work out its density without melting it down.

To carry out the experiment, Archimedes had two crowns made, each of the same weight as the one produced by the suspect goldsmith. One was composed of pure gold, the other of silver. He filled a container to the brim with water and immersed the golden crown, measuring the exact amount of liquid it displaced. He then did the same with the silver crown. As he had expected, the result

Archimedes in his bath
A 16th-century illustration of the sage just before his 'Eureka!' moment features spheres and crowns of different sizes – and, no doubt, of varying densities.

was different, indicating that gold and silver have different densities. By this time Archimedes was close to answering Hiero's conundrum. The final step lay in submerging the goldsmith's crown in the container. Sure enough, it displaced less water than the silver crown but more than the gold. As Hiero had suspected, the crown was therefore made of an amalgam of gold and silver, and the fraudster was duly punished.

The laws of floating bodies

As for Archimedes, he used his experiment with the crowns to deduce the principle that still bears his name – that any object wholly or partly immersed in a fluid is uplifted by a force equal to the weight of the fluid displaced.

Within that formula lies the explanation as to why some objects float in water while others sink to the bottom. Archimedes explained his conclusions in his *Treatise on Floating Bodies*, thereby launching the study of hydrostatics. The fundamental laws that he outlined addressed one of the main preoccupations of Greek shipbuilders of his day: how to construct a boat that sits well in the water even when heavily loaded.

Archimedes was far from the first person to investigate this problem – it had presented itself in empirical fashion ever since humans first set out to build boats. But he was one of the earliest to seek to resolve the problem by the application of mathematics.

A great mathematician

The great inventor was first and foremost a mathematician. It is thought that he may have studied in his youth under Euclid in Alexandria, but even if not, he was certainly brought up on the master's work. His own contribution to pure mathematics would be considerable: he provided a formula for calculating the value of π (pi) – the ratio of a

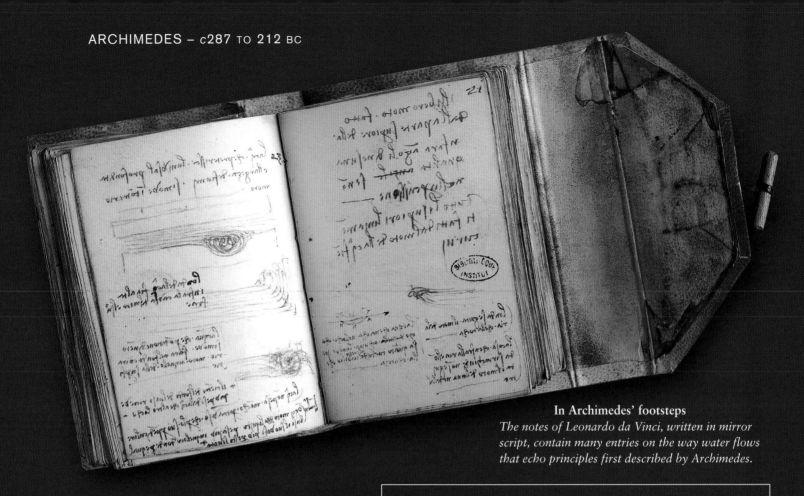

In Archimedes' footsteps
The notes of Leonardo da Vinci, written in mirror script, contain many entries on the way water flows that echo principles first described by Archimedes.

circle's circumference to its diameter – by measuring the perimeter of two polygons, one inscribed within the circle and the other circumscribed around its edge. He also improved the theory of real numbers by proposing the so-called Archimedean property, by which certain algebraic structures (known as Archimedean groups) contain no infinitely large or small elements – so-called non-zero infinitesimals. In some respects this work heralded modern integral calculus.

Archimedes also devoted much time to the study of area, discovering formulae to measure the spaces enclosed by parabolas, cylinders and spheres. He did important work on spirals and on the volumes of sections of cones and other so-called surfaces of revolution – those created by a curve rotating around a straight line.

Following in the footsteps of Euclid, on moving back to Syracuse Archimedes constantly sought to link pure maths with practical technology. He drew up axioms from his observations of the physical world that led him through thought processes, like the one that came to him in his bath, to formulate fundamental mathematical laws.

Putting theory into practice

In this way Archimedes established a vital link between the calculation of area and the centre of gravity of plane surfaces. He propounded the Law of the Lever, which states that magnitudes are in equilibrium at distances

AN ENDURING LEGACY

Most of the surviving writings ascribed to Archimedes are theoretical works. With the exception of the *Treatise on Floating Bodies,* which founded the science of hydrostatics, and a work called *On the Equilibrium of Planes,* concerned with statics, the rest are mathematical, either dealing with geometry, as in *On the Measurement of Circles,* or with algebra and infinity, as in *The Sand Reckoner.* Archimedes' ideas were so much in advance of their time that mathematicians in Europe continued to rely on them up to and even after the Renaissance. His books were set texts for the study of maths until the 17th century and he remains an important point of reference in physics to this day.

The power of leverage
A 17th-century edition of Vitruvius's De Architectura *illustrates the principle spelled out by Archimedes.*

proportional to their weights. He established a new branch of physics in the shape of statics, concerned with forces operating on solid bodies at rest. He found mathematical formulae to explain the operation of levers, working out that the ratio of the force applied to raise a load is equal to the inverse ratio of the distances of force and load from the lever's pivot. 'Give me a lever and a place to stand', he claimed, 'and I will move the Earth.'

Putting his ideas into practice, Archimedes combined levers with mobile pulleys – also his own invention, according to tradition – to develop powerful cranes and hoists. He also made improvements to the screwpump to improve the Archimedes' screw, a hydraulic device that still bears his name (see page 61).

A violent death

Even in his lifetime Archimedes enjoyed an extraordinarily high reputation as a practical inventor. Entrusted by Hiero with the task of defending Syracuse from the Romans, he succeeded by his ingenuity in greatly prolonging the city's resistance. After a bitter siege lasting more than two years, the Roman general Marcellus finally forced a way into the city, helped by the treachery of a pro-Roman party within its walls. Orders had been given for Archimedes to be spared, but a soldier who was unaware of the command came upon the

savant as he sat rapt in contemplation of mathematical figures traced on the ground. When Archimedes failed to heed the order to surrender, the soldier ran him through. In conformity with Archimedes' wishes, a sphere inscribed within a cylinder was set upon his tomb.

Magnificent defence
Archimedes' giant hook-and-pulley may have been too heavy for a warship, but it was possibly used to upend Roman ships during their siege of Syracuse.

ARCHIMEDES AT WAR

Archimedes flung himself energetically into the defence of Syracuse when a Roman fleet besieged the city in 214 BC. He devised various machines to hurl missiles at the enemy vessels, including the ballistas that worked like giant crossbows: a cross-string was drawn back by means of a winch, then released, propelling darts or cannonball-shaped stones up a ramp inclined at an angle of 45 degrees in the direction of the enemy. In practice, the ballista's range was limited to about 150m. Catapults, which had been in use since about 450 BC, were more effective. They employed the leverage obtained from twisted ropes to crank back a giant arm, throwing heavier stones as far as 1,000m.

Tradition also claims that Archimedes employed a heat-ray device against the enemy vessels thronging the Gulf of Syracuse. Adapting a basic principle of optics to military use, it is said that he stationed numerous mirrors in a wide parabola on the city walls in order to catch the sun's rays and reflect them onto a precise spot on one of the ships. According to a Greek named

Didymus, 'the sailors hardly had time to realise that the vessels had caught fire…'. Later writers have poured scorn on the claim, on the grounds that the tools available lacked the precision to achieve the described effect, even in the hands of a genius like Archimedes. In contrast, ballistic weapons like those employed by the Greeks at Syracuse continued in use until the 16th century.

Burning-glass
Modern engineers doubt whether the heat-ray weapon attributed to Archimedes could have achieved the claimed effects

The parachute *c*150 BC

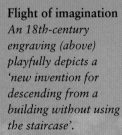

The watching crowd must have been tense and expectant one day in August 1812 when André Jacques Garnerin parachuted in free fall from a balloon floating 1,000m above the town of Clermont-Ferrand in central France. The descent turned out to be successful and the intrepid inventor lived to tell the tale

Daring though it was, Garnerin's exploit was not quite as unprecedented as the spectators may have thought at the time. Parachutes already had a surprisingly long history. Scholars have discovered that as early as the 2nd century BC acrobats in China demonstrated their courage by jumping off high towers hanging from giant umbrellas. Thereafter the parachute was 'reinvented' more than once in different eras and places, to be first seriously tested in the 18th century and constantly improved by engineers and enthusiasts since.

A law of physics

The idea behind parachuting, and the principle that sustains it, stems from Isaac Newton's Third Law of Motion, which states that for every action (or force) in nature there is an equal and opposite reaction. This concept of reciprocity implies that the air resistance operating on a large, umbrella-shaped device might be sufficient to sustain the weight of a human body in free fall, the degree of resistance (and hence the carrying

Unexpected anachronism
A 16th-century Book of Hours *by René d'Anjou has a marginal illustration of an angel apparently paragliding.*

Flight of imagination
An 18th-century engraving (above) playfully depicts a 'new invention for descending from a building without using the staircase'.

LEONARDO DA VINCI'S PARACHUTE

The great Renaissance genius left to posterity some 6,000 technical drawings, a number of which involved flying machines. Among them was a prototype parachute, pyramid-shaped on a square base and made from linen stretched over a rigid frame. Accompanying notes explained the principles involved. Leonardo employed mirror-writing in his notebooks to keep his thoughts private.

Parachutes for flying
The large fabric wings of paragliders (above), divided to form a row of cells, are designed to give the pilot maximum glide-time.

capacity) being proportional to the surface area exposed to the air and to the square of the speed of the fall. This formula also explains why parachutists actually slow down in the course of the drop.

Putting these ideas into practice proved far from simple. Leonardo da Vinci addressed the problem at the turn of the 15th and 16th centuries. Perhaps inspired by Leonardo's work, the Venice-based Croatian Faust Vrančić successfully launched himself from a tower in 1617, suspended beneath a wooden frame that supported a stretched square of cloth. In June 2000 a British parachutist, Adrian Nicholas, successfully jumped from a height of 3,000m using a replica of Leonardo's design, thus confirming its practicality despite the fact that experts from America's National Aeronautics and Space Administration (NASA) had earlier concluded that it was too unstable to work.

A new wave of interest

In the late 18th century many individuals were bitten by the urge to fly, and the concept of parachuting started attracting widespread attention. In 1783 a Frenchman, Louis-Sébastien Lenormand, jumped from a tree with the aid of a pair of modified umbrellas; later in the same year he employed a refined version of the device to perform a public jump in the southern city of Montpellier. Besides crossing the Channel in a balloon, the aeronaut Jean-Pierre Blanchard (1753–1809) also

Pioneering leap
In 1797 André Jacques Garnerin (above) parachuted from a balloon floating above the Parc Monceau in Paris.

experimented with parachutes, while another pioneer balloonist, Joseph de Montgolfier, had the idea of attaching one to a wickerwork balloon gondola.

As for Garnerin, the Clermond-Ferrand demonstration was not his first attempt. He had jumped from a balloon 1,500m above Paris 15 years earlier, landing with the aid of a circular chute 11m across. The idea for his invention came to him while he was being held by the British as a prisoner of war during the Revolutionary Wars in 1794.

Multiple descents

The birth of aviation gave fresh impetus to the evolution of parachuting. The first person to risk a parachute jump from an aeroplane was probably the American Grant Morton in 1911, who completed his successful descent over California. Parachutes were used by German pilots during the First World War, but they played a far more significant part in the Second World War, both by saving the lives of airmen and by enabling troops to land behind enemy lines. In the post-war years design improvements have introduced better steering and compensation for wind drift. After self-opening parachutes reduced the risk of malfunction, parachuting became a sport – and one with an enviably low accident rate. Today's enthusiasts of skydiving, paragliding, BASE jumping and the other thrills now available to the adventurous owe a debt of gratitude to those brave pioneers who first leaped into thin air, not knowing for sure whether they would survive or were going to their death.

The gimbal *c*140 BC

The upper reaches of Chinese society enjoyed a refined and comfortable lifestyle in the 2nd century BC. Otherwise they would hardly have had a need for the gimbal, which was invented to prevent the contents of perfume-burners from spilling onto cushions. To achieve this end, the burner was placed at the centre of linked concentric circles. When the chain from which the device hung was shaken, the circles absorbed the motion, while the central burner remained upright.

In Europe the first gimbal is credited to the Greek inventor Philo of Byzantium (280–220 BC), who described an eight-sided inkpot with an opening on each side. A pen can be dipped and inked no matter which face is on top, yet the ink never runs out through the holes on the other sides. The inkwell was suspended at the centre of a series of concentric metal rings, which remain stationary no matter which way the pot is turned. Centuries later a gimbal-like device featured in the 13th-century notebook of master-builder Villard de Honnecourt, who travelled around France recording details of the great Gothic cathedrals. Among the devices he sketched was an untippable heater, designed to warm the frigid interiors of churches.

THE UNIVERSAL JOINT

In 1676 the scientist and astronomer Robert Hooke borrowed ideas from the gimbal to devise Hooke's joint, a connection between two rigid rods that allows them to move freely in different directions. The Hooke arrangement involves two hinges set at a 90-degree angle to one another and connected by a cross-shaft. It can be used in drive-shafts to transmit rotary motion. The first use of the term 'universal joint' dates to the late 19th century, and this alternative name subsequently entered common parlance when the joints began to be widely used in the manufacture of the first Ford automobiles. Also occasionally known as a Cardan joint, it features to this day in most vehicle transmission systems.

The invention did not attract much attention until the mid-16th century and the time of Girolamo Cardano (1501–1576). This Milanese jack-of-all-trades, a doctor as well as a mathematician, physicist and philosopher, described the gimbal and how it functioned in a work entitled *De subtilitate rerum* ('On the subtlety of things'). In the years that followed gimbals were put to serious use as mountings for ship's compasses, the design counteracting the ship's pitch and roll and so permitting captains and helmsmen to find true north even in stormy weather.

Gimbals on paper and in practice
The sketches made by Villard de Honnecourt in the 1230s (above) are now preserved in the Bibliothèque Nationale in Paris. The ball-shaped perfume-burner (top left) can roll in any direction without scattering its contents.

The magic lantern c120 BC

Illusions can exert a fascination, as the Chinese realised early in their history when they devised the art of shadow puppetry. They also quickly learned that the spectacle is more seductive, even magical, if real images are substituted for the shadows. Projecting images required a sophisticated understanding of optics – and there, too, the Chinese were ahead of the game.

Ancient Chinese texts describe an instrument recognisably similar to later magic lanterns. Silhouettes – either cut out or traced on a translucent backing – were placed upside down in front of a light source in a darkened room and projected in magnified form with the aid either of a rudimentary lens or a simple hole serving to concentrate the light rays. An entertainer named Shao Ong is said to have used such a device to entertain Emperor Wu, who ruled from 141 to 87 BC.

Cinema's ancestor

In the West some authorities have attributed the invention of the magic lantern to the English scholar Roger Bacon in the 13th century. Bacon certainly took an interest in optics, and almost everyone who studies that field inevitably turns their mind at some point to the projection of images. The revival of optics in the 17th century led another English savant, John Bate, to describe a magic lantern in 1634. Yet the lanterns only became widely known in the early 19th century, when several individuals

contributed to their development, among them the physicist Michael Faraday.

The latter part of the 19th century was the golden age of magic lanterns when many technological developments improved their performance. Over the course of a few decades rudimentary devices gave way to elaborate systems with multiple lenses and reflecting mirrors that allowed the focus to be adjusted and the light source brightened or dimmed. The next step was the illusion of movement, created with the aid of discs or loops of repeated, slightly altered images, as in a zoetrope. The Lumière Brothers, who staged the first public screenings of projected motion pictures in Paris in 1895, were partly inspired by such devices. Beyond that point the magic lantern's evolution became part of a bigger story: the history of cinema.

Precursor of the film projector
Made in 1810, this large magic lantern (above) was used to project so-called 'fantasmagorias' – cut-out paper silhouettes. Such exhibitions were popular in Europe at the time (top left).

Shadow plays
Puppets on sticks are a long-standing tradition in story-telling in Indonesia.

THE ASTROLABE – c150 BC
The star-catcher

Invented by a Greek astronomer, the astrolabe was used to chart the position of stars in the sky and was the instrument that scholars needed to calculate the movement of the heavenly bodies. Adopted by the Arabs and then later by the West, it served for almost two millennia as an essential tool for astronomers.

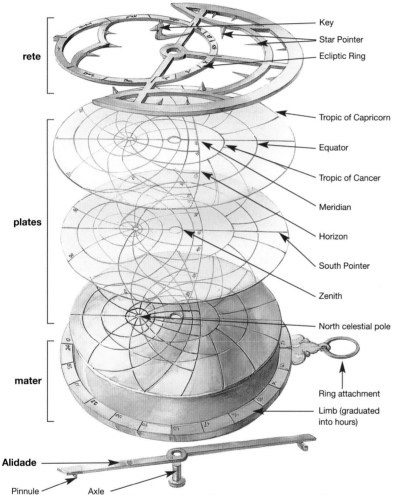

- Key
- Star Pointer
- Ecliptic Ring

rete

- Tropic of Capricorn
- Equator
- Tropic of Cancer
- Meridian
- Horizon
- South Pointer
- Zenith
- North celestial pole

plates

mater

- Ring attachment
- Limb (graduated into hours)

Alidade

Pinnule (or Sight)   Axle

Key to an astrolabe
The diagram above illustrates the parts of a typical astrolabe.

Essential tool
An astrolabe (top right) that once belonged to Ahmed ibn Khalaf, who lived in Baghdad c850.

The Greek astronomer who invented the astrolabe more than 2,000 years ago had a clear idea in mind: to note the position of the stars. Simple enough in its basic concept – a moveable pointer attached to a marked frame – the instrument's original purpose was to fix the altitude of heavenly bodies, that is, their elevation above the horizon measured in degrees. The word 'astrolabe' derives from the Greek *astron* ('star') and *lambanein* ('to take' or 'to catch'), loosely translatable as 'capturing the position of a star'.

A Mediterranean invention

There is some uncertainty as to who came up with the idea. One candidate is Apollonius of Perga (*c*262–180 BC), an astronomer and geometer who devoted much of his time to writing an eight-volume treatise on conic sections. Apollonius's mathematical training would certainly have given him the knowledge required to map the three-dimensional heavenly sphere in two dimensions around a

---

A PLANISPHERIC ASTROLABE

An astrolabe had four main components. The *mater* supported the other three, to which it was linked by a central axle, capped at the top by a key. Gradations round the rim (officially known as the limb) served as hour-markers allowing the user to tell the time. The *plates* were stacked on one side of the mater; the markings on each plate were specific to a given latitude, so fresh plates were required when the latitude changed. The engraved lines represented projections of lines corresponding to the movements of heavenly bodies. The *alidade* was a moveable rod fixed

to the mater's reverse side and equipped with two sights known as pinnules. When the astrolabe was held vertically by the ring attached to the mater, the user could aim the sights at a star and read off its elevation above the horizon in degrees, as shown on a scale round the rim. The *rete* was a cut-out plate that rotated over the latitude plates below it. As the rete turned, its star pointers marked out the position of the stars they represented against the celestial sphere as engraved on the plate beneath. The ecliptic ring was the Sun's path through the sky as seen from Earth.

hypothetical celestial equator – the precondition for devising a simple astrolabe. But there is no direct evidence that Apollonius actually performed the task.

In about 150 BC Hipparchus, the greatest of Greek astronomers, took up where Apollonius had left off. He developed a system for fixing the position of stars that in practice performed the function of an astrolabe. Yet the first recorded mention of the instrument post-dates Hipparchus by almost three centuries. Ptolemy mentions a rudimentary device in his *Tetrabiblos*, a treatise on astrology.

The earliest surviving description of a planispheric astrolabe (like the one shown on the left) comes from the writings of John Philoponus, who lived in Alexandria from about AD 490 to 570. His model incorporated several engraved plates that between them supplied the means for working out a considerable amount of information from the position of the stars in the sky: the time of day, the hour of sunrise, the direction of the cardinal points, and so on. Philoponus' treatise also provided the solution to eleven astronomical problems that he resolved with the aid of an astrolabe.

The astronomer's tool of choice

From the 8th century the astrolabe was taken up by scientists in the Islamic world, who had learned of the instrument from translations of Greek texts. In addition to its other uses, it was employed to determine the hours of prayer and also to work out the direction of Mecca on desert journeys. From Baghdad and Antioch, knowledge of the device reached Persia and northwest India. By that time astrolabes, which were often made of precious metals, had themselves become works of art.

HIPPARCHUS, ASTRONOMER OF GENIUS

Much of what is known of Hipparchus (190–120 BC) has come down through the writings of Ptolemy (AD 92–168), who described the earlier astronomer's works in the *Almagest*. Considered the greatest of classical astronomers, Hipparchus compiled a catalogue of 1,080 stars that was unprecedented in its day. He pioneered the use of trigonometric principles, permitting him to calculate with reasonable accuracy both the distance from the Earth to the Moon and the Moon's diameter. He drew up theoretical models to explain the different duration of the seasons in terms of the movements of the two heavenly bodies, and also used his mathematical skills in combination with his knowledge of the Sun and Moon to develop one of the first reliable systems for predicting eclipses.

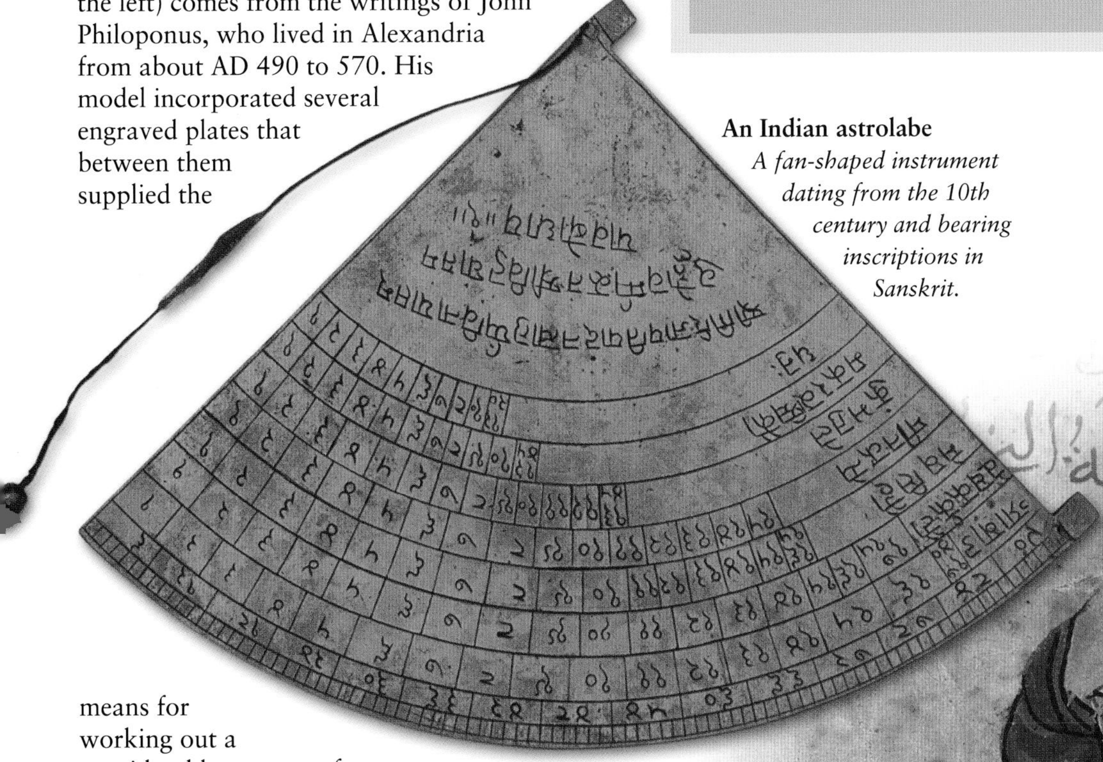

An Indian astrolabe
A fan-shaped instrument dating from the 10th century and bearing inscriptions in Sanskrit.

Studying the stars
A 13th-century Turkish astronomer (below) demonstrates the use of an astrolabe to his pupils.

Sailors at sea
A 15th-century French miniature painting (above) shows a mariner navigating the Indian Ocean with the help of an astrolabe.

Following in the wake of the Greeks and Arabs, European astronomers enthusiastically adopted the astrolabe when it arrived sometime around the year 1000, having come by way of Christian monasteries in the north of Spain. Gerbert of Aurillac, who in 999 became Pope Sylvester II, did much to spread knowledge of the device. Another keen proponent was Hermann of Reichenau (1013–1054), who took his name from the Benedictine abbey on Lake Constance where he spent most of his life. In two treatises on the uses of the astrolabe, Hermann provided solutions for 21 separate astronomical problems. The first astrolabes to reach western Europe retained their original Arabic inscriptions alongside added Latin tags. Because of this certain stars, such as Altair, Vega and Deneb, are still known in the West by the names originally given them by earlier Arab astronomers.

MORE THAN A POET

The English poet Geoffrey Chaucer (c1340–1400), most famous for his *Canterbury Tales*, was also a philosopher, diplomat and writer about science. In his unfinished but widely disseminated *Treatise on the Astrolabe*, he used the year 1391 for his calculations, which scholars believe to be the year in which the son of his friend died. This was the 'Little Lewis' to whom the treatise is dedicated. An astrolabe dating to 1326 and matching the one Chaucer describes can be seen in the British Museum.

Decline and fall

Until the 17th century the astrolabe remained an essential tool for astronomers, but from the turn of the 18th century fewer were made. Modern prismatic astrolabes have little in common with the historic instrument that shares their name, other than the similar function of measuring the elevation of stars. Meanwhile, traditional astrolabes fell out of use and became museum pieces.

Even so, the instrument has a secure place in the history of astronomy, having enjoyed more than two millennia of uninterrupted use. By comparison the refracting telescope, invented by Galileo a mere four centuries ago, remains something of a novelty.

How the Earth became round

Almost 2,000 years before Copernicus and Galileo, Greek astronomers were the first to propose that the Earth is a spherical body floating in space. The most audacious thinkers even suggested that it revolved around the Sun. They also managed to estimate its circumference, along with that of the Moon.

It is hard to put a date on the birth of astronomy, for people have been fascinated by the heavens ever since they first cast their eyes up to observe the stars, the phases of the Moon, or solar and lunar eclipses. Some people see megaliths such as Stonehenge, built in stages over the course of the 3rd and 2nd millennia BC, as spectacular evidence of ancient fascination with celestial questions. Researchers claim that the great stones were oriented to the rising and setting of the Moon and Sun, specifically being aligned with sunrise at the time of the summer solstice. Some have suggested that the stone circle might have been used to predict eclipses; others argue that, given the importance accorded to the heavens in early times as the home of the gods, the site's builders had a religious motivation. In the absence of any written documentation, the mystery of the monument's ancient function is likely to remain unresolved.

Yet observation of the skies also served more down-to-earth purposes. With the coming of agriculture some 10,000 years ago, our ancestors needed to predict the weather and seasons to judge the right time for sowing seeds. That preoccupation led to the creation of the first calendars, drawn up by the Mesopotamians – who cultivated the lands on the banks of the Tigris and Euphrates rivers in what is now Iraq – and by their counterparts along the Nile in Egypt. These astronomers based their calculations on the motions of the heavenly bodies and how their recurrent patterns correlated with the annual cycle of the seasons. They showed less interest in the nature of the stars and planets themselves. That field would be left for the Greeks.

Centre stage
Sin, the Assyrian moon god, is carved at the centre of this 8th-century-BC stele. Most early civilisations peopled the heavens with gods and goddesses.

Ancient circle
Stonehenge dates back to the 3rd millennium BC. Some argue that it had an astronomical purpose, perhaps as a reference point for observing the motions of heavenly bodies.

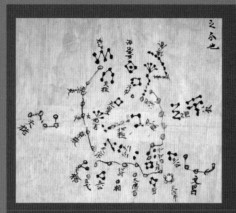

ASTRONOMY IN CHINA

Chinese astronomers were in general more concerned with observing the night sky than with devising theories to explain the workings of the heavens. They had long studied solar and lunar eclipses; engraved bones, used for divination, have been found referring to a lunar eclipse that occurred in 1361 BC and to a solar one in 1216 BC. Chinese scholars worked out the length of the year to 365 and a quarter days, and calculated the so-called Metonic cycle of 19 years, after which the phases of the Moon recur on the same day of the month. They also knew of the lunar eclipse cycle of 135 lunar months.

Thales and the nature of matter

In about 580 BC, Ionia on the Asian shore of the Aegean Sea was experiencing something of a golden age. Governed by an oligarchy of rich citizens, the city of Miletus was one of the most prosperous in the entire Mediterranean world. In this era of free thought and trade, a military engineer called Thales came forward to revolutionise the history of thought.

Thales brought back an impressive stock of knowledge from travels in Egypt, where he reportedly diverted the course of a river that was blocking the advance of the army of King Croesus of Lydia. He is also said to have predicted a solar eclipse, although history does not record how. In Egypt he mastered local surveying techniques used to divide large areas of land into small plots with dimensions that could be measured with the aid of wooden posts linked by ropes. He followed this up by providing a theoretical explanation for the pragmatic methods of the Egyptians, thus laying the foundations of geometry. Yet none of these achievements would be considered Thales' greatest legacy. For that one must begin with a question that Thales set out to answer: what exactly is the world made of?

No civilisation up to that date had questioned the fundamental nature of matter. The Greeks saw the Earth's landmass as a flat disc – one that had rested since the beginning of time on solid pillars – that was surrounded by a fast-flowing ring of ocean and covered by the solid vault of the heavens. The stars travelled across the vault from east to west, only to be carried back to their starting point by the ocean stream.

Thales preferred to visualise the universe as a bubble of air surrounded on all sides by water. For him the Earth we live on was a flat surface at the base of the bubble, while the stars moved around its top. There was no place in his schema for the ocean stream, nor, more importantly, for the world-supporting pillars.

The Round Earth theory

In the mid 6th century BC the Persian invasions spelled the end of the glory days for Miletus. It was the turn of the Greek cities of

Reading the stars
A copy of Ptolemy's Tetrabiblos, *prepared in about AD 820 (below), includes this illustration of a Byzantine zodiac. For many centuries astronomy and astrology were intimately linked.*

Star map
Views of the heavens changed relatively little from the time of the Romans, when this map was made (above), up to the Renaissance.

PLATO'S AXIS OF THE UNIVERSE

Summing up the astronomical knowledge of his day, Plato postulated in the *Timaeus* that the universe turned on a central axis in the same manner as the Earth, which he held to be the 'guardian and artificer of night and day'. From these words he clearly believed that the Earth controlled the motions of the Sun, a view that astronomers would hasten to debunk in later centuries.

Sicily and southern Italy to shelter the best minds of the epoch. Of these, none was more distinguished than Pythagoras (*c*570–490 BC) who, like Thales, had learned his maths partly from the Egyptians. Pythagoras founded a school at Croton in Calabria with the aim of examining all natural phenomena from a mathematical perspective. For him and his pupils the Earth was spherical, a theory supported by everyday observations. After all, when a ship approaches port, the sail usually appears over the horizon before the hull.

A logical next step was to ask whether, if the Earth was round, the same might not also be true of the Sun and Moon. For subsequent generations of Greeks the Earth was a sphere within a sphere, with the heavenly bodies attached to the outer rim. Seven were known at the time – the Sun and Moon plus the five planets visible to the naked eye, namely Mercury, Venus, Mars, Jupiter and Saturn.

Parmenides (*c*515–440 BC) was the first person to suggest an explanation for the phases of the Moon. A disciple of Pythagoras who later challenged some of his master's views, Parmenides came to believe that the Moon owed its light to the Sun and that its changing form came from the fact that it was viewed from the Earth at different angles. The idea was revolutionary, not least because previously no-one had ever considered the nocturnal dark to be merely an absence of light, viewing it rather as an opaque mist that somehow emanated from the Sun at nightfall.

The Pythagoreans themselves then came up with a novel suggestion of their own. Philolaus, who was born early in the 5th century BC, proposed a system in which the Sun, Moon, Earth and five planets, along with a mysterious body he named the Counter-Earth, all revolved on spheres arranged around a central fire. In such a system our planet ceased to be the centre of the universe.

A fresh round of speculation

Athenian thought in the 5th century BC showed a bent for abstract reasoning that forced direct observation to take a back seat. The result was some theoretical models of the universe that owed little to physical reality. The great Plato (*c*427–347 BC) twice addressed astronomical problems, first in the *Republic* and then in the *Timaeus*, postulating that the five known planets and the Sun were all equidistant from the Earth and that they all moved around the heavens in the same direction with the exception of Mars, which for some unknown reason seemed to go the other way. In the *Laws* he insisted on the

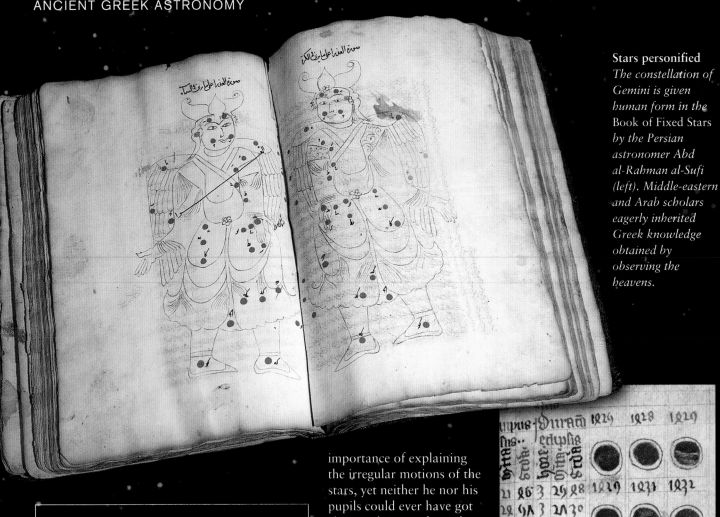

Stars personified
The constellation of Gemini is given human form in the Book of Fixed Stars *by the Persian astronomer Abd al-Rahman al-Sufi (left). Middle-eastern and Arab scholars eagerly inherited Greek knowledge obtained by observing the heavens.*

IN THE ANCIENT WORLD

Following its introduction by Anaximander (610–546 BC), the Babylonian gnomon or sun pointer was used to measure time in ancient Greece. A vertical rod planted in the ground cast a shadow; by observing its length and angle in relation to the shadow at midday it was possible to work out the passing of the hours. At night people studied the motions of the heavenly bodies. Egyptian astronomers aligned two plumb-lines against the Pole Star to create a north–south meridian, then observed the passage of certain stars across the line. A less complicated method involved observing the flow of liquid in a water clock, which in Egypt had twelve marks to indicate the passing of hours.

To take sightings, the Greeks used an alidade – a movable rod attached to a fixed point and equipped with sights referred to as 'pinnules' (see page 72). Used together, two alidades became a compass, allowing the user to measure the angle between stars or between a star and the horizon. Armillary spheres were criss-crossed by circles and provided with sights, allowing astronomers to orient them against the position of stars in the sky. But such devices only became available in the latter part of the Hellenistic era.

importance of explaining the irregular motions of the stars, yet neither he nor his pupils could ever have got close to doing so from the premises he proposed. In the *Timaeus*, for example, he argued that the Sun's diameter could be no more than one-eighth that of the Earth. Fortunately, pupils do not always slavishly follow their master's teachings, no matter how highly regarded. In the *Epinomis* – once mistakenly attributed to Plato himself but now generally accepted as the work of Philip of Opus, a member of his Academy – the author correctly insisted that all the stars, and particularly the Sun, were in fact much bigger than the Earth.

Surveyors of the skies

By the 3rd century BC Alexandria, the capital of Hellenistic Egypt, was a magnet for the best Greek minds. Sheltered there from the assaults of northern barbarians, scholars had access to the finest library of the ancient world. In about 230 BC Eratosthenes, who held the position of head librarian, set himself a novel challenge: to measure the circumference of the Earth. The task was a daunting one, not least because

Solar eclipses
Ancient Greek astronomers found ways of calculating the timing of solar eclipses. The knowledge resurfaced in Europe during the Renaissance, as this English calendar of 1408 suggests (above).

the Greek world was limited at the time to the Mediterranean Basin. To estimate the total size of a globe that remained largely unknown, Eratosthenes turned to geometry. On a visit to Syene (today's Aswan, in southern Egypt), which lay on the Tropic of Cancer, he noted that at noon on the summer solstice sunlight lit up the bottom of a well – something that did not happen at Alexandria. He realised that, to calculate the fraction of the Earth's circumference that lay between the two cities, he had only to measure the number of degrees off the vertical of sunlight hitting Alexandria on the same day of the year. From his measurements he reckoned that the angle represented roughly one-fiftieth of the Earth's circumference. So he then multiplied the distance between the cities by a factor of 50, thereby achieving a good approximation of the actual figure (fractionally over 40,000km, or almost 25,000 miles). Eratosthenes's achievement was remarkable, given the paucity of evidence available at the time. It was only achieved through the power of pure mathematical reasoning.

Aristarchus of Samos (c310–250 BC) also turned to geometry to address the related question of the Moon's circumference, seeking to calculate it from information provided by eclipses. He measured the time that the Moon took to travel through the shadow cast by the Earth, which he imagined (wrongly) as a cylinder with a circumference equal to that of the Earth all along its length. He used the figure thus obtained to conclude that the Moon was roughly one-third the size of the Earth. The actual figure is just over a quarter and Aristarchus' error is explained by the fact that the Earth's shadow tapers into a cone, not a cylinder.

Aristarchus also challenged the consensus that the Earth lay at the centre of the universe. Taking up an idea originally floated a century earlier by Heraclides of Pontus, he affirmed that the apparent motion of the planets in the sky could be convincingly accounted for if the Earth, like the five other known planets, was in fact revolving around the Sun. His words counted for little set against the opinion of Aristotle and the theory, although correct, failed to make much

Astrolabes in medieval manuscript
In a copy of the fanciful Travels of Sir John Mandeville, *written in the 14th century, the monks of Mount Athos observe the night sky*

Celestial model
The collection of circles in this medieval armillary sphere represent the movements of the Sun, Moon, planets and stars around the Earth,

Instead, Hipparchus and later Ptolemy (AD 92–168) devoted great ingenuity to devising geometric models to explain and predict the apparent motions of the planets, seeking thereby to reinforce the established Aristotelian worldview. The models, of course, are wrong as seen from the viewpoint of modern astronomy, yet they still stand as monuments to the sophistication of several centuries of Greek mathematics. When Ptolemy published his *Almagest*, summing up his vision of the universe, he gave Greek astronomy its most comprehensive masterwork – one that would continue to be used as an essential reference for the next 1,400 years. Ptolemy's views would not be seriously challenged until the 16th century, when a young Polish astronomer finally called the master's ideas into question. His name was

CONCRETE – *c*100 BC

Artificial stone that helped to build an ancient empire

The discovery of how to make concrete allowed the Romans, from the 1st century BC, to construct vast monuments, bridges and aqueducts that paved the way for imperial greatness. The secret of making the material was subsequently lost, and many centuries would pass before its unique properties were once more appreciated.

Archetypal amphitheatre
The Colosseum (below right) seated about 80,000 spectators in its day. It is still one of Rome's leading attractions.

The use of concrete and mortar was not entirely new when the Romans made their breakthrough discovery. The Greeks had used a lime mortar to bond masonry blocks, while an aqueduct built in Iraq in about 700 BC incorporated a layer of concrete made from a mixture of lime, water, sand and gravel. The material was permeable, however, so it had to be covered with a protective layer of asphalt. What was new about Roman concrete, first devised by workmen at Pozzuoli near Naples, was the creation of an effective cement from water and lime that served as a binding agent for both mortar and concrete.

A new building block

The novel ingredient in the mix was volcanic sand, which in Naples was readily available thanks to the proximity of Mount Vesuvius. The result was remarkable. The oxides of iron, aluminium and silicon contained in the solidified lava, called pozzolana, marked a step improvement on the mortar known to the Greeks: the end product closely resembled modern cement. Recent studies have shown that the Romans also made artificial pozzolana by calcining (heating) crushed bricks or volcanic rocks. Cheap to make, easily moulded and able to bear heavy loads, the resultant cement played an essential part in Roman architectural projects.

Roman cement was hydraulic, which is to say that it hardened on coming into contact with water. The concrete made from it was thus suitable for use in humid or marshy

THE ROMAN COLOSSEUM

In AD 72 Emperor Vespasian gave the orders for the building of the Colosseum, the greatest of all Roman amphitheatres; eight years later, his son Titus officially opened the building. For its construction, the builders used the best materials to hand, including wood, stone, brick, marble – and concrete. According to modern estimates, about 6,000 tonnes of concrete went into the building, much of it for the foundations, considered an extraordinary feat of engineering at the time. With outer walls rising 48.5m high and a central arena the size of 13 tennis courts, the Colosseum was state of the art in public buildings of its day. The heart of the complex was below ground level in a two-storey basement called the hypogeum, which contained cages for the wild animals and the technical equipment needed for the shows. The arena was linked to a nearby aqueduct so that it could be flooded for naval spectacles.

to be rediscovered until the 18th century. In 1756 an English civil engineer named John Smeaton invented what he called 'hydraulic lime' – a form of cement that set as hard as rock – but he made relatively little use of it. From 1817 the chemical properties of cement became known thanks to the work of a French engineer, Louis Vicat. Seven years later, the manufacturer Joseph Aspdin took out a patent on Portland cement, taking the name from the oolitic limestone found on the Isle of Portland in Dorset, which it resembled. His son William improved the product from 1841 by increasing the limestone content and burning the mixture much harder. From that time on, the cement business went literally from strength to strength.

Constant improvements

Cement played a crucial part in the development of reinforced concrete, invented in the 1860s by a French gardener named Joseph Monier. Monier discovered that he could make constructions watertight by covering them with a layer of Portland cement, which had the added advantage of preventing the iron rods that reinforced the concrete from rusting. Thereafter reinforced concrete was increasingly used in the construction of tunnels and underpasses and for the roofs and terraces of apartment blocks.

In Roman times concrete, like stone, had been employed for its strength and resistance to pressure. Following Monier's invention and then the development of pre-stressed concrete in 1886, it also became a material of choice for its flexibility. The extra tensile strength made it possible to extend the span of bridges and beams, opening up a whole new range of possibilities for architects.

Concrete marvel
The Pantheon in Rome, which was built in the 1st century AD, still boasts the world's largest unreinforced concrete dome, 43.3m in height and width (left).

New possibilities
The Port Boulevard Bridge in Miami (below) is one of many contemporary structures that could not have existed in its present form without the invention of reinforced and pre-stressed concrete.

environments. It could be used to build port facilities, as at Caesarea in Palestine, where blocks weighing more than 50 tonnes were sunk into the sea. It was also employed in the foundations of bridges and aqueducts.

Lost and found

The architects of the Middle Ages and the Renaissance preferred stone to concrete, and over the ages the Roman recipe was lost, not

A new medium for the written word

By inventing paper, the Chinese produced a handy, lightweight and inexpensive surface for writing. Use of the new material spread rapidly in a civilisation that relied heavily on the written word, opening the way for another major development that followed in its wake – printing.

According to Chinese history, at some unspecified date Cai Lun, a functionary at the Han court, was given a challenging commission. He was told to find a new medium for writing that would be cheaper than silk – in use for the past four centuries – and easier to handle than the strips of bamboo held together by silk strips that were used for everyday communications. After years of research Lun came up with a material that fitted the bill: leaves of paper made by soaking and crushing vegetable fibres plus scraps of silk, old fishing nets and mulberry bark.

Old and new

So the story goes, as told in the official records of the Han Dynasty. Archaeology has added further detail with the discovery, in a 1st-century-BC tomb in northern China, of fragments of paper made from hemp fibre.

It now seems likely that Cai Lun took as his starting point an old method of recycling hemp and jute known in China from the 5th or even 6th century BC. The fabric was soaked in water and beaten, then put out to dry on mats. Cai Lun's contribution lay on the one hand in using fresh materials and on the other in drawing official attention to the process, for his work was well received at the imperial court. Perhaps, too, he was the first person to think of using the recycled material for writing. The earliest scraps of paper have been dated to 110 or 109 BC, which ties in with the traditional story. In practice, the Chinese were developing a new application of the papyrus-making technique long since used in ancient Egypt.

Spreading the word

Reflecting past practice, the Chinese ideogram for 'paper' is an adaptation of the one representing 'silk'. After Cai Lun's time, the

Hemp scroll
A Chinese scroll dating from the 7th century AD is made of ten separate leaves of heavy-duty hemp paper, glued end to end and coated with wax.

PARCHMENT

In the 2nd century BC the city of Pergamon in Asia Minor became one of the leading cultural centres in the Greek world. Its library even rivalled the library at Alexandria, a fact that so displeased Egypt's ruling Ptolemies that they cut off supplies of papyrus to the city. In response, Pergamon's artisans took to using a fresh material, namely sheep or goat skins, and parchment was born, taking its name from the Greek *Pergamenos*, meaning 'of Pergamon'. More accurately, parchment was reborn, for animal skins had been used for writing in earlier days both in Egypt and Mesopotamia, where separate words existed to denote 'tablet scribes' and 'skin scribes'.

The parchment-makers of Pergamon quickly developed techniques for producing high-quality merchandise. The skins were washed and then bathed in limewater for up to two weeks, before being carefully scraped to remove any trace of flesh or hair. Then they went back into the lime solution, emerging to be stretched for a final scraping. The resulting surface was smooth enough to support the marks made by a goose quill, which was easier to write with than a reed stylus. Better still, sheets of parchment could be folded, stacked and sewn together to form codices, early books that were less cumbersome to read than papyrus scrolls.

MEXICAN BARK PAPER

Paper was in use in the great city of Teotihuacan in the 5th century AD, and bark paper was still in demand in central Mexico a millennium later. The Aztecs, who by that time ruled the land, produced three different varieties made respectively from palm fibres, agave leaves and fig-tree bark. In each case the procedure was the same. The vegetable fibres were soaked in water to soften them – a process known as retting – and the resulting pulp was pounded with stones and finally polished with ears of maize. The leaves of the paper were made up into codices that could be as much as 33m long, folded concertina fashion.

principal materials used were first mulberry bark and then, from the 10th century, bamboo and rice straw, all of which were cheaper and easier to come by than silk, even in the form of cast-off clothes.

The manufacturing process was relatively straightforward. First the fibres were cleaned and left to soak in vats filled with water. Once they had softened they were boiled to a pulp. To remove the pulp from the water, paper-makers used a fine-mesh bamboo screen in a reusable wooden frame, which was plunged into the vat rather in the way that a cook uses a sieve or skimmer: the pulp was caught by the mesh while the water drained through the trellis. The layer of softened fibre left on the screen still contained a great deal of water, so next it was pressed between two layers of wood or stone to squeeze out the liquid. The leaves thus obtained were then spread out on a warm surface to dry.

With state encouragement paper-making spread rapidly across the imperial lands. By the end of the 2nd century leaves of paper were in use even in the most remote regions owing allegiance to the Han.

Multiple uses

Despite their readiness to adopt the new writing material, Chinese scribes at first remained loyal to long-established habits in the way they used it, with the leaves glued end-to-end to form scrolls in the manner of the old rolls of silk. There were obvious inconveniences in this arrangement and before long practical considerations dictated that lengths of paper should instead be folded concertina fashion to form books.

As paper quality improved, calligraphers and painters adopted the new medium. The paper they chose for their most cherished works was made from long fibres of rice straw and hackberry wood that was thought to be particularly durable. Meanwhile, cheaper forms of paper were being put to a variety

Pre-Columbian illumination
The precious Codex Fejérváry-Mayer, a manuscript of illustrations from Aztec times, survived the Spanish conquest of Mexico intact and unharmed.

The cutting of bamboo *to make paper is shown in a 19th-century illustration (above). The same procedures are still in use today; the workman (left) is binding stems that have already been cut to size.*

Making paper
An 18th-century engraving (right) shows workmen preparing wood pulp, then drying and pressing the leaves. The Chinese Uighur artisan below is using similar techniques to prepare paper from mulberry wood.

of uses, from making fans and parasols to lanterns and kites. Coated with oil, it was substantial enough for the manufacture of blinds and the screens used to partition rooms. It could even be used for personal hygiene in the form of paper towels and tissues.

Printing money

By the 8th century the everyday use of paper for writing had laid the groundwork for another major technological advance: the development of woodblock printing, allowing text and pictures to be printed leaf by leaf. Before the century was out Chinese tea-merchants were using promissory notes called *feiqian*, literally 'flying money'. In 1023 the state took control of paper currency; the first banknotes authorised by the imperial government were issued in Sichuan province the following year. The Mongol khans who ruled China under the Yuan Dynasty made considerable use of paper money, much to the astonishment of the few Western travellers such as Marco Polo, who visited the Yuan court in the late 13th century.

Paper trail

By that time paper had finally reached Europe. Knowledge of its use had reached Korea and Japan around the year AD 600, but another 150 years passed before paper-making became known to the Arab world. It arrived after the Battle of Talas in 751, when Islamic forces took Chinese prisoners who were prepared to trade their knowledge of the technology – previously preserved as a precious secret – in exchange for their lives.

Paper reached Baghdad by way of Samarkand in Central Asia in 793, when Harun al-Rashid, the fifth Abbasid caliph, decreed its use in place of the more expensive parchment and papyrus. Soon the capital had its own paper factory. Heading west, the technology reached Damascus in Syria, Tripoli in what is now Lebanon and Egypt early in the 10th century, then Morocco by 1100. By that time papyrus had virtually fallen into disuse in the Islamic world.

The Berbers finally brought paper to Europe via the lands in Andalusia ruled by their Almohad Dynasty. The first paper-making manufactory on the continent was established at Jativa, in the Valencia region, in 1150. The technology spread into France and Italy, but Britain's first paper mill was not established until 1495. Until that time all paper was imported from Europe.

Technical improvements

In Europe, as in the Arab lands, paper was made primarily from rags, obtained by breaking down linen and hemp fibres. The process was first mechanised at Jativa with the introduction of a papermill to speed the drying of the slurry. In Italy the industry established itself in 1276 at Fabriano, near Ancona, where the artisans introduced the process known as sizing – soaking in a bath of animal glue, so the leaves absorbed less ink. Thereafter they provided the continent with a vastly improved product for much of the next century. The Fabriano workers were also the first to use watermarks, making it possible to identify the maker of a particular batch of paper.

Even so, it took the invention and spread of printing for paper finally to supplant parchment across the continent. And printing would never have happened without the invention of paper in the first instance.

Chinese prayers
Today, as in the past, people leave prayers written on small pieces of paper in Chinese temples (above).

Ancient method
A young Laotian boy irons sheets of virgin paper in the traditional manner.

VELLUM

Usually made from the skin of stillborn calfs, vellum was a parchment of superior quality that could support writing on both sides, like surfaces prepared from sheepskin but not from goatskin. Another advantage of vellum was that it was not greedy for ink – the pigments used for manuscript illumination did not soak into it, keeping their original colours better. The finest miniatures of the Middle Ages were painted on this costly material.

Hero and the school of Alexandria

Hero of Alexandria was a mathematician and engineer of genius, one of the most illustrious exponents of Hellenistic science. His inventions – which included the first steam engine, automatons, an odometer for measuring distance travelled and a precursor of the surveyor's theodolite – rank him among the all-time giants of science.

Deadly weapon
Hero's design for a crossbow (below) used tightly twisted cords to power the shot. The result was a fearsome bow that is a scaled-down version of technology used in the cheiroballistra, a Roman siege engine.

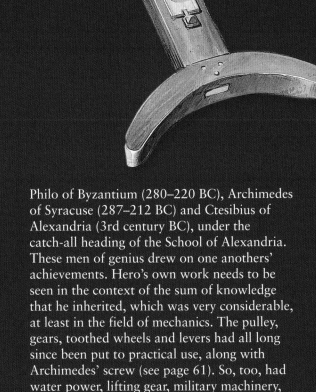

Mechanical figure
Automatons, like this singing canary made in 1895, had counterparts in 1st-century-BC Greece.

It seems likely that Hero was working in Alexandria when the city was captured by the Romans in the 1st century BC. The text of a treaty dealing with agricultural matters, credited to one Columella and dated to the year 62 BC, borrowed exact phrases that Hero used to describe ways of calculating land areas and that have been preserved to this day. *Automata*, Hero's work on automatons, was published in Latin translation in the 16th century and widely read. His other important works include the *Belopoeica*, a treatise on machines of war, the *Cheiroballistra*, concerning catapults and siege engines, and the *Pneumatica*, exploring air, steam and water power. These are thought to have been produced as lecture notes for students.

Inheriting a tradition

Hero's place in the history of science is special partly because of the diversity of his interests – he was an innovator in many fields, ranging from mechanics and physics to maths and astronomy. Then there was the sheer fascination that his work inspired when it was rediscovered during the Renaissance. There was also the fact that Hero represented the end of a long line of physicians and mathematicians who had shown great ingenuity in putting their ideas into practice. As master of the Alexandria school, Hero was able both to transmit and add to this precious heritage of knowledge.

Historians have loosely grouped the most illustrious of Hero's predecessors, including Archytas of Tarentum (4th century BC),

Philo of Byzantium (280–220 BC), Archimedes of Syracuse (287–212 BC) and Ctesibius of Alexandria (3rd century BC), under the catch-all heading of the School of Alexandria. These men of genius drew on one anothers' achievements. Hero's own work needs to be seen in the context of the sum of knowledge that he inherited, which was very considerable, at least in the field of mechanics. The pulley, gears, toothed wheels and levers had all long since been put to practical use, along with Archimedes' screw (see page 61). So, too, had water power, lifting gear, military machinery, measuring equipment and, perhaps most surprisingly of all, automatons.

Technology at play

Many of the inventions credited to Hero, as well as various improvements he made to earlier discoveries, apparently came about in

the context of his teaching duties, for he established and ran a school of engineering. Others seem to have been devised purely for the fun of it, with no other object than to astonish and entertain. The Greeks of Hero's day had a taste for ingeniously operated automatons. Rather like the masterpieces that medieval craftsmen were expected to produce to display their technical proficiency, these machines were not so much intended to have immediate practical applications (except in the military field) as to demonstrate the skill of the maker. Economically, the need for machinery was a low priority in an age when the practice of slavery provided a cheap and abundant labour force.

There was a steady flow of remarkable automaton creations. In the 4th century BC

Archytas constructed an artificial bird with jointed wings that seemed to fly when a jet of steam was directed at it. A century later a Greek military engineer, known as Philo of Athens to distinguish him from his Byzantine namesake, devised ingenious contraptions that included a robot wine-server employing a ballcock mechanism operated by steam or water pressure. Hero's reputation as a wonder-worker derived largely from the fantastic machines that he described in the *Automata*, including trumpets that sounded unbidden, temple doors that opened automatically when fire was burned on an altar and articulated statues that seemed to drink, along with many other devices that drew on his mastery of mechanics and physics.

Modern impression
A 1st-century-BC workshop in Alexandria, as interpreted by a modern Western artist.

CITY OF KNOWLEDGE

Alexandria was founded in 332 BC by Alexander the Great. In the reign of Ptolemy I Soter, who ruled from 323 to 283 BC, it became an intellectual powerhouse influencing the entire Hellenistic world. Ptolemy himself helped to make the city a hub of scholarship by founding a museum and laboratories as well as the city's famous library. His successors in the Ptolemaic Dynasty followed his example, going out of their way to attract the finest scholars to the city. Drawing on the heritage of both Greece and Egypt, the School of Alexandria eventually boasted in its ranks such luminaries as Archimedes, Euclid and Callimachus in the field of mathematics, Erasistratus, Herophilus and Dioscorides in the natural sciences and medicine. The astronomer Aristarchus of Samos, the geographer Strabo and the great Claudius Ptolemaeus (Ptolemy) all contributed to the city's fame.

Robotic wine-server
The automatic wine pourer worked through the weight of water shifting between different levels to move levers. As the right arm holding the pitcher was raised, the left arm with the bowl was lowered. Once the bowl was full, an overflow system reversed the process, restoring the statue to its original position.

Water organs and fire engines

In the field of hydraulics Hero followed in the path of Ctesibius of Alexandria (see panel opposite). Ctesibius was responsible for several remarkable machines that worked by water pressure, among them the hydraulis or water organ, which used the weight of water to compress air inside a metal dome; when the keys were pressed valves holding back the air opened, releasing it into the pipes and causing them to sound.

Valves were at the heart of another of Ctesibius's inventions, the clepsydra or water-clock. This featured a water reservoir equipped with a float connected by a rod to a statue holding a pointer. Water flowing through a pipe into the tank raised the float and with it the attached pointer, whose position against a calibrated cylinder marked the passage of time.

Ctesibius's pump, as described by Philo of Byzantium in the 2nd century BC, was rather more complex. It consisted of two cylinders immersed in water, each equipped with a piston linked to a lever and joined by a conduit through which water could escape. The motion was continuous: when the lever was lowered to compress the piston in Cylinder B, water was forced into Cylinder A, then the process was reversed when the lever operating Piston A was lowered, causing Piston B to rise again. The system was used into modern times in ships' pumps and – connected to a horse-drawn cistern – on early fire engines.

Hero's fountain

Hero made improvements to Ctesibius' pump by providing the outlet pipe with a pivoting head that made it possible to direct the jet of water. He also came up with ingenious water-powered inventions of his own, displaying a similar mastery of the laws of hydraulics. Famously he described a device still generally known as Hero's fountain. This requires an open basin and two separate, airtight water containers set at different levels, with only the upper one filled with water. Water poured into the basin runs through a tube into the lower

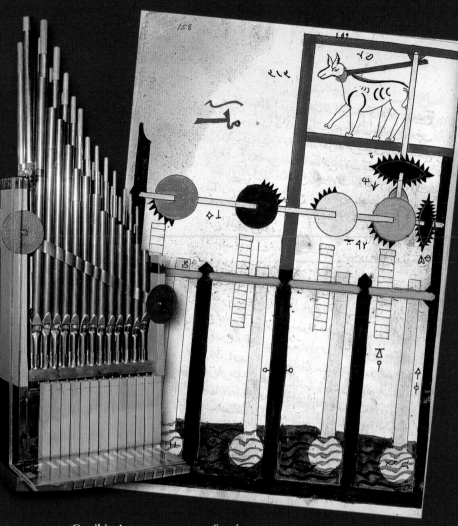

Ctesibius' organ
A reconstruction of Ctesibius' original design (above left), as described in the 3rd century BC.

Suction pump
A Seljuk manuscript of the 13th century (above) explains a sakia, or suction pump, in schematic form. The Arabs and later the Turks preserved Classical Greek manuscripts and later went on to develop the ideas contained in them.

The fire engine
Another of Ctesibius' inventions inspired this 19th-century mechanical toy.

CTESIBIUS, INVENTOR OF GENIUS

One of the founding fathers of the School of Alexandria, Ctesibius was responsible for inventions as varied as the hydraulis or water organ, a gun operated by compressed air, a spring-loaded crossbow and an astronomical clock. If one quality could be said to characterise Ctesibius' work, it would be pertinacity, for he went to unusual lengths in following through his ideas.

The power of compressed air

Noting a remarkable property of air – its elasticity – Ctesibius set about making cylinders that he first polished smooth inside and then soldered shut, leaving just room for the rod of a piston to pass through one end. He quickly established that, even if the cylinders were airtight, he could still force the piston down thanks to the compressibility of the air inside. The next step was to fasten the piston with a safety catch. When this was released, the air expanded abruptly, driving it rapidly upward.

By this experiment Ctesibius neatly demonstrated the existence of energy contained within compressed air. He then proceeded to apply the principle by taking a tube to serve as the compression chamber and placing a projectile within it. When the air pressure was released, the energy generated was powerful enough to fire the missile for 150m. Without knowing it, he had invented the prototype of the air gun. At the time his invention had little impact. His contemporaries remained committed to bladed weapons and his discovery was regarded as little more than a scientific curiosity until the 16th century, when the first 'wind-guns' were manufactured in Germany. Even then, few people availed themselves of the new weapons, and it was only with the invention of the air rifle in the 19th century that Ctesibius' principle became widely applied.

Another of Ctesibius' inventions was a development of the crossbow, replacing the twisted cords used up to that date with steel springs compressed to their utmost and held in place with a safety catch. When the trigger was released, the stored energy propelled the arrow forward. This device, too, failed to catch on and was no more successful at the time than the air gun.

Counting the hours

Water clocks, which divided up the passage of time into twelve hours, were not equipped to cope with the fact that in summer the nights are shorter than the days, while in winter the opposite is true. Ctesibius set about addressing this problem. He devised a clepsydra with a float in the receptacle, which rose with the incoming water. A vertical rod attached to the float meshed with a rotating drum on which lines were drawn representing the months and the unequal hours. This drum was regularly adjusted so the pointer travelled over a distance representing the hours of daylight for that particular day. In one version of the clock Ctesibius incorporated a statue bearing a pointer that moved over a scale on the drum that was marked off in hours.

Chinese astronomical clock
A marvellous piece of Chinese time-keeping machinery, dating from the 11th century and standing almost 10m high. It was powered by a water wheel connected to a geared drive shaft of the type pioneered by the engineers of the School of Alexandria.

container, forcing out the air through a second tube into the upper receptacle and causing the water within it to spout into the basin. Much to the delight of watching spectators, the fountain could continue in operation for several hours, until all the water in the upper reservoir was finally used up. Another amazing invention was based on the principle of linked vessels, using a ballcock mechanism rather like that found in modern flush toilets, to regulate the flow of water into a tank that was fed from an external water source through a tube.

Putting gears to work

In the field of mechanics Hero built on the work of Archimedes. He devised a triple-pulley crane and pursued his predecessor's work on raising heavy loads with the aid of levers and pulleys, coming up with the concept of the auger elevator. The *baroulkos*, which Hero described in his *Mechanica* ('Mechanics'), employed the principle now used in car gearboxes, by which motion transferred between two cogs of different dimensions increases the torque of the larger one proportionally to the degree to which its rotation is slowed. The device consisted of a metal casing containing a series of gear wheels, each bigger than the one before. By turning a crank, the operator activated a screw that set the gears in motion. The final gear wheel was connected to a winch with a rope attached, making it possible for a single person to raise weights that were otherwise well beyond the lifting capacity of any human or draught animal. The same principle of economy of effort underlay other inventions by Hero, notably a lifting machine that employed a worm-drive mechanism activated by a capstan connected to a toothed wheel, whose axle operated a lifting cable.

Assimilated knowledge,
In a design for a self-filling bath (above), a 13th-century Seljuk manuscript displays ideas borrowed from Hero and other scientists of the School of Alexandria

Living tradition
The hydraulic ideas of the Greek engineers are given a contemporary twist in the sculpture fountain outside Paris's Pompidou Centre, designed by the artists Niki de Saint-Phalle and Jean Tinguely.

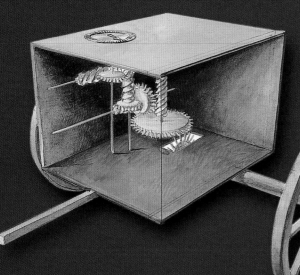

The transfer of energy
Gearing (left) was one of the most valuable discoveries made by the engineers of the School of Alexandria.

Experiments with steam

Hero's explorations of energy included research into what happened to air and water under pressure. His work in this field helped to inspire an invention that many centuries later would lead to the development of turbines. This was the aeolipile, or wind-ball, the world's first steam engine. Often simply referred to as Hero's engine, this took the form of a hollow sphere, equipped with bent nozzles, connected by piping to a sealed cauldron filled with water. When the water was heated, steam rose through pipes into the sphere and was vented through the nozzles, causing the sphere to rotate. Hero seems to have conceived the engine simply as a toy.

Perhaps a more impressive invention at the time, at least to a crowd of unwitting spectators, was an invisible mechanism for opening temple doors. This ingenious device required an impressive array of technological know-how and considerable knowledge of the

The first steam engine
In Hero's aeolipile, steam generated in the covered cauldron rose through hollow tubes into the upper sphere, where it was vented through two bent nozzles, causing the sphere to rotate.

TAKING MEASUREMENTS

Like many of his predecessors in the School of Alexandria, Hero was not just an inventor and engineer but also a mathematician and student of geometry. In that capacity he wrote the *Metrica*, a work dedicated to measurement in which he drew not only on Greek theory but also on the empirical methods used by the Egyptians and Babylonians to calculate surfaces and volumes. He used the work of Euclid in dividing up areas and volumes into two parts with a given relationship to one another, and went on to devise a method for working out the square root of numbers. Similarly, he explored optics in a work called the *Catoptrica*, which dealt with the problems posed by the reflection of light in mirrors, establishing the basic law of reflection that light takes the shortest path between two points. In his *Dioptra* he provided a detailed description of the machine of that name, an ancestor of the theodolite that was invented by Hipparchus in the 2nd century BC, which could be used to measure angles. Hero improved on Hipparchus's model by adding a water level, enabling users to establish a true horizontal.

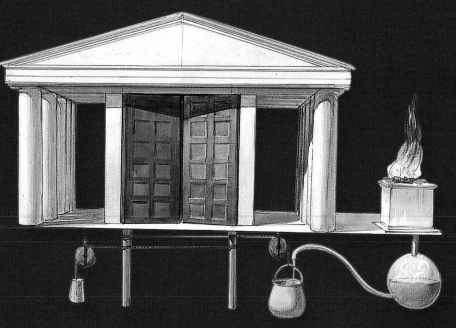

Automatic doors *Hero sought to amaze onlookers with his proposal to use steam power generated by an altar flame to open temple doors by means of unseen pulleys.*

in reverse, closing the doors. Thanks to Hero's mechanism, technology could achieve a seeming miracle – and one that could be repeated at regular intervals to a set timetable.

Machines of war

Like most of the ancient world's scientists and engineers, Hero also turned his attention to military technology, which was the field that offered both the best financial rewards and the greatest opportunity for the practical application of new discoveries. He left behind detailed descriptions of a range of powerful ballistic weapons that had been developed by a line of illustrious engineers, beginning with Archimedes in the 4th century BC. The projectile power of these catapults gradually increased as a result of technological advances in the field of levers and pulleys.

Hero himself came up with a version of the cheiroballistra, a giant portable crossbow designed to fire arrows over a great distance. Metalworking had advanced by his day, and it was a sign of the times that his device was no longer made entirely of wood like its predecessors. Instead it incorporated many parts made out of iron or steel, notably the springs used to stretch the bowstring that served to fire the missiles.

laws of physics. The heat generated by a ceremonial fire burning on an altar was used to warm water in an underground sphere. This generated steam, which set in motion a system of cords and pulleys that between them created the rotary motion required to open the temple doors, as if by magic, before the astonished gaze of worshippers. Putting out the sacred flame caused the whole process to be repeated

LEONARDO, GREECE'S HEIR

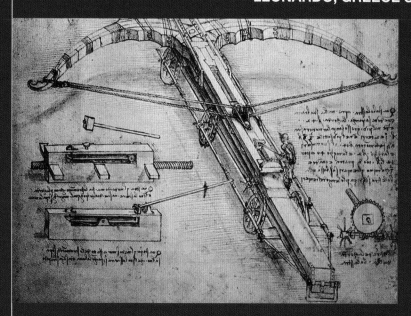

Spring-action crossbow *Leonardo drew inspiration from the inventions of the School of Alexandria, while adding improvements of his own devising.*

Hero's writings and the rest of the corpus of knowledge assembled by the School of Alexandria, spread first across the Roman Empire and then later, after Alexandria fell to the armies of Caliph Omar in the mid 7th century, through the Islamic world. The Arabs safeguarded the body of scientific knowledge as Europe plunged into the Dark Ages, and it was through them that ancient learning reached the West in the course of the Middle Ages. With the dawn of the Renaissance the intellectual climate in Europe once more encouraged free inquiry. No-one expressed its spirit better than Leonardo da Vinci (1452–1519), who immersed himself in the writings of the ancients to find inspiration for his own original researches. Like Hero, he became fascinated with the power of heat, imagining a machine consisting of a propeller turned by hot air operating a system of gears and pulleys to rotate a spit for roasting meat. In 18th-century England John Barber took out a patent on the first turbine. His invention, featuring a paddle wheel turned by heated gases, was recognisably a successor to Hero's aeolipile.

The crane *c*45 BC

In AD 72 work began in Rome on a building whose name alone suggests its massive size. The Colosseum was the biggest amphitheatre ever constructed by the Romans, providing seating for some 80,000 spectators. Unlike the builders of the Egyptian pyramids, however, who only had ramps to raise the stone blocks as their constructions grew higher, the Roman masons had access to a recently invented tool to help with their work – the crane.

First described by Vitruvius in the 1st century BC, the Roman crane had its origins in an A-shaped frame held firm by retaining cords. The lifting gear consisted of a rope that passed via a pulley to a winch turned by a handle. Rudimentary though it might have seemed to modern eyes, this contraption was

used by the Romans to build the great monuments that dot the once-imperial lands.

By the time of the European Renaissance, cranes had grown rather more sophisticated. As shown in the notebooks of Leonardo da Vinci and the engineer Francesco di Giorgio Martini, by this time the crane consisted of a wooden frame equipped with a jib. The lifting machinery was activated either by a treadmill, operated by an individual treading paddles set into its outer rim, or by a treadwheel, with the operator or operators at work inside. Machines of this type had been employed on major construction sites since the Middle Ages. They were also used in great ports to load and unload ships; the best pivoted on a vertical axle, permitting them to be swung into place.

Increased lifting power

The first cast-iron cranes were made early in the 19th century. Soon new models were being constructed of iron girders, then from steel, always with the aim of increasing their lifting power so they could handle ever-heavier loads – for instance, locomotives. For all their hoists and gears, these cranes still relied on human labour to operate them. Later in the century, steam power would boost their lifting capacity tenfold.

Other innovations also contributed to creating the implements that have built the modern world. Tower cranes came into use to cope with high-rise buildings. Telescopic cranes were developed to reach places where traditional ones could not go. The oil pressure in the hydraulic system that operates telescopic cranes is maintained by huge diesel engines, giving today's machines a capacity that the engineers of antiquity could only dream of.

Weight-lifting
A 16th-century illustration shows a pivoting crane lifting supplies to the top of a tower.

Moving power
Mobile cranes were sometimes operated by treadwheels even in Roman times (left). The modern equivalents below are mounted on tracks.

The arch – the secret of spanning distances

From the earliest times people have needed to find ways to cross rivers and gorges. Bridges date back into prehistory, but the Romans were the first to master the construction of arches, enabling builders to project roads and aqueducts across wide spans. The principles involved in arch construction would also benefit other fields of architecture.

Bridge span
From early times, arches were a crucial element in bridge design. Considerable technological expertise was required to build structures like the Arch of Tiberius at Pompeii (right), erected in the 1st century AD, or the bridge at Cyrrhus in Syria (below), which Roman engineers constructed in the following century.

Generally people hate wasted effort. To avoid making repeated detours around natural obstacles, they sought from early times to find a way to cross the barriers, if necessary by building permanent structures over them. The first bridges were probably built across streams in the Stone Age using the cantilever method: facing pillars would have been erected on each bank and planks weighted with stones set on top of them, joining in mid-stream. The point where the planks met would then have needed reinforcement to carry loads, which even so could never have been very heavy.

In about 3500 BC, when bricks first came into use, Mesopotamian engineers learned how to make mortared, flat-topped corbel arches, constructed with overlapping layers finally meeting in the middle, and possibly also radial arches, with bricks rising in a semi-circle from the pillars, each held in place by its neighbours. These techniques were used to build the first bridges over the Tigris and Euphrates.

Architectural breakthrough

Technology continued to evolve over the next 3,000 years. Stone replaced brick as the principal construction material in Egypt, Babylon, Greece and Persia. The Etruscans, who came from the east to occupy northern Italy in the 8th century BC, brought with them knowledge of corbelling and other oriental building methods. None of their bridges have survived; the arches were probably crudely finished and, like their counterparts elsewhere in the ancient world, must eventually have collapsed.

Three centuries later the Romans put Etruscan technical know-how to use to carry through a hugely ambitious project: the geographical unification of Italy by means of a road network stretching the entire length of the peninsula. Bridges would play a vital part in carrying the highways across the many rivers and watercourses that cut across the face of the landscape. In the course of their work, Rome's engineers had to resolve two basic problems posed by the semi-circular arch. Using masonry, as they regularly did to build edifices capable of bearing the weight of a marching legion complete with supply wagons, they first had to find a way of supporting the massive dressed stones in the course of construction before the bridge was completed.

THE MARCH OF THE ARCH

Before arches there were lintels, formed simply by setting a block of stone across two pillars. The first proper arches were built in the Middle East in the 4th millennium BC as elements in underground water systems. The main problem facing the builders lay in supporting the stones or bricks during construction. The largest arch erected before Roman times was at Thebes in Egypt, in a grain warehouse dating from about 1400 BC; researchers think that the builders put up a temporary brick structure to hold it in place as they worked. Assyrian buildings at Khorsabad, dating from around 700 BC, also featured brick vaults.

The Romans were the first to master the true arch, built of stone without mortar. They used wedge-shaped blocks called voussoirs supported on a temporary wooden frame that was removed after the central keystone had been added. Thereafter the voussoirs were held in place by their own weight.

In the early Middle Ages the semi-circular Roman arch was given a new lease of life as a key feature of the appropriately named Romanesque style of architecture. Builders in the Ile-de-France region around Paris then discovered that pointed arches reduced the horizontal thrust on the supporting walls, allowing them to build higher. The result was the Gothic style, which developed from the mid 12th century. Cathedrals thereafter became lighter in appearance, seeming to soar heavenward. Pointed arches were never popular for bridge-building, partly because the abutments on each bank could usually sustain the considerable sideways thrust. Renaissance architects subsequently rejected the Gothic arch in favour of flattened, semi-elliptical designs, no longer forming a complete semi-circle as in Roman times. By reducing the angle of inclination, they did away with the need for humpback bridges.

No mortar was strong enough to hold the blocks in place. Instead, the engineers devised the technique known as centring, which involved building a wooden arch to hold the stones until all were in place, after which the timber framework was dismantled. Similar techniques are used to this day by builders working with stone or concrete blocks.

The other difficulty involved planting pillars in the beds of rivers without them being swept away by the current. The answer in this case lay in constructing watertight caissons that could be filled with concrete, a recent Roman invention. These retaining structures provided a dry environment for workers to construct the load-bearing pillars that supported the arches.

Water course
The Roman acqueduct at Segovia in Spain brought water from a source 18km (11 miles) outside the city.

Bridge over the River Danube
A bridge built to plans drawn up by Apollodorus of Damascus in AD 103 also features on Trajan's Column (above), which was erected in Rome to celebrate the Emperor's victories in the Dacian Wars.

Bridges to empire

Technological prowess enabled the Romans to take on increasingly ambitious projects. In 55 BC, at Julius Caesar's orders, engineers constructed a crossing over the River Rhine in just ten days. Rome itself was endowed with eight separate stone bridges over the Tiber in the two decades after 220 BC, giving citizens of every district easy access across the river.

Improved communications benefited trade, but that was never the main priority. The first concern was always military. By making it easier for the legions to move freely to where they were needed, the bridges played an important part in the spread of empire.

Ensuring water supplies

The Roman genius for civil engineering encouraged administrators to attempt ever greater feats, and none was more important than providing water supplies to cities. Roman urban centres used huge amounts of water, and in many cases water from wells and rainwater runoff proved insufficient to meet demand. It then became necessary to find some reliable source in the surrounding countryside.

To bring the precious liquid into the city centres the Romans built aqueducts – canalised channels, some covered but others not, that sometimes stretched for 100km (60 miles) or more. The first step was to find a source of water to tap, set in firm ground with an underlay of solid rock. Next, watercourses were prepared to direct the water into a basin that served as a reservoir. From there the liquid flowed into conduits known as leats, which marked the start of the aqueduct proper. Usually, leats sloped downhill below ground level, following the natural contours of the landscape, for subterranean conduits were relatively cheap and easy to construct.

Sometimes, the lie of the land forced the engineers to seek other solutions. To traverse uneven terrain or cross a shallow dip, the conduit might have to emerge above ground and continue its course along the top of a specially built wall. Usually this would be equipped with masonry arches if the land fell away by more than 2m. Viaducts were constructed that bridged entire valleys.

Giant endeavours

In the 1st century AD an aqueduct was built to carry water 50km (30 miles) from Uzès to Nimes in southern France. The main obstacle across the route was the valley of the River Gardon. To overcome it, engineers built the Pont du Gard, the highest viaduct in all the Roman lands, rising 49m above the valley bottom. It took the form of three bridges, one on top of another, made up of semi-circular arches – six on the bottom tier, eleven on the second and 35 on the top level. The span of

Long-lasting structure
An aqueduct (left) built in the 2nd century AD at Uthina (now Oudna in Tunisia) on the orders of the Emperor Hadrian remained in use for 1,500 years.

Roman achievement
The Pont du Gard (below), carrying an aqueduct that served the Roman city of Nemausus (today's Nimes in southern France), is now listed as a UNESCO World Heritage Site.

STRANGE RECIPE

In Roman times conduits were made watertight by covering them with a mixture of lime, sand and brick rubble that was then coated with a substance known as malthe. This curious concoction, also sometimes called mineral pitch, was said by Pliny the Elder to consist of 'lime ground up with pork fat and figs ... No coating has greater protective powers; it sets harder than stone.'

the arches on the lower levels was exceptional, measuring 19.2m and 24.5m respectively; a typical size at the time was just 5.5m.

In practice viaducts were rarely used to cross drops of more than 45m because of the risk of collapse. In these situations the Romans adopted another solution – one involving an inverted siphon arrangement. The water was carried down the valley side in a conduit constructed of lengths of lead piping. It then travelled across the river atop a low-level arched bridge, after which the pressure generated by the drop was sufficient to carry it up the other side to resume its normal flow at the top. Such constructions were costly and often sprang leaks, so relatively few were built. Even so, in the 1st century AD the city of Lugdunum (modern Lyon), capital of the province of Gallia Lugdunensis, depended on eight of these arrangements for its water supply.

At the end of its journey the water spurted out into a reservoir, to be distributed across the city through a network of clay or lead pipes. Vitruvius in his *De Architectura* ('On Architecture') described an arrangement by which the water in the reservoir was then divided for distribution into three smaller basins; one fed the city's fountains where most citizens went for their water, another the public baths and the third the few private residences that could afford a direct connection to the mains supply.

Suspension bridges ancient and modern
Two Nepalese sherpas cross a bridge in the Himalayas (below) built on the same principle as the Pont de Normandie (above), which opened in 1995 and is considered a masterpiece of modern bridge design.

THE SUSPENSION PRINCIPLE

People in many parts of the world today still use rope bridges to cross narrow gorges and rapids, just as their ancestors did thousands of years ago. These structures are erected on an entirely different principle to arched bridges, as they rely for their support on cables stretched between pillars set at either side of the space to be traversed. The first recorded suspension bridge is mentioned in a Chinese chronicle of about 25 BC; it crossed a gorge in the Himalayas and had a span of 15m. Chinese records also mention a bridge 120m long in the same region that was built 400 years later of bamboo, still used in China for scaffolding. From China, too, came the first structure to be suspended on metal chains; erected in 1638, it stretched for 45m. Thomas Telford was responsible for the first steel suspension bridge in the West, constructed in 1826 across the Menai Strait separating Anglesey from the North Wales coast. Some 200m long, it still remains in use today.

After the Romans

Few improvements on Roman bridge-building technology were made in the course of the Middle Ages. Some structures were erected incorporating buildings, like the Ponte Vecchio in Florence or the old London Bridge. The situation changed with the Renaissance, when Italy witnessed the development of 'surbased' bridges – low-lying structures with flattened arches whose rise above water level was less than half the span. The Rialto Bridge spanning the Grand Canal in Venice is a fine example.

In France the creation of the Corps of Bridges and Roads in 1716 signalled a general improvement in construction techniques. By the end of the century, England had taken the lead in the use of new materials with the construction of the famous Iron Bridge in the Shropshire town of the same name, situated in the heartland of the Industrial Revolution. Completed in 1779, the central span stretched for 30.5m between two masonry abutments, which supported the bridge at each end.

For more than two millennia, engineers have sought to build ever more ambitious structures by combining new materials with the most up-to-date technologies. The Humber Bridge, the world's longest suspension bridge when it opened in 1981, not only included towers of reinforced concrete – a material first employed in the early 20th century – but also incorporated hollow steel boxes rather than solid steel trusses, enabling the bridge to withstand winds of up to 130km (80 miles) per hour. The year 2004 saw the opening of the Rio-Antirio Bridge across the Gulf of Corinth in Greece, featuring the world's longest cable-held suspended deck.

Bridge of buildings
The Ponte Vecchio, the oldest bridge in Florence, is still lined with houses and shops just as it was when it was built in the mid 14th century. At the time the practice was common.

SETTING RECORDS

The biggest bridge in the Roman world is thought to have been one built across the River Danube at the behest of the Emperor Trajan incorporating 20 spans each 38m across. The first flattened-arch bridge built in the West was the Ponte Vecchio in Florence. The Charles Bridge in Prague, also dating from the 14th century, was the longest bridge built in the Middle Ages, stretching for 516m. Gustav Eiffel, the builder of the Eiffel Tower, took the record for the largest iron arch bridge when he built the Garabit Viaduct, stretching for 552m across the Truyère River in France's Massif Central. His claim was superseded by the construction of the Sydney Harbour Bridge in Australia in 1932. Twenty-five years later San Francisco's Golden Gate Bridge became the world's biggest suspension bridge, with a length of 1,281m. The world's first iron bridge was designed by Thomas Farnolls Pritchard and built over the River Severn by Abraham Darby III between 1777 and 1779. Remarkably, it was constructed entirely using mortice and tenon joints – no screws or rivets were used.

Highways across empires

Wheeled vehicles created a demand for roads – paths suitable for travellers on foot or on horseback had to be transformed into thoroughfares that could carry heavy carts and carriages. Linked in networks, these highways stand to this day as monuments to the power and dynamism of the states that built them.

Roman roads did not always make for comfortable travelling – at any rate not if one can judge from the letters of the philosopher Seneca. Travelling to Naples in AD 55, he complained bitterly about the mud and the dust. Yet the discomfort did not stop people from using the highways; traffic criss-crossed the Roman Empire, the first to create a true road network across its lands.

The first great civilisations, Egypt and Mesopotamia, grew up beside navigable rivers and initially relied on water transport as their main channel of communication. With the invention of the wheel in the 4th millennium BC, wagons and war chariots started using whatever tracks were available. The first roads were built to connect places that could not be reached by sea or river. So in Egypt, in the latter years of the Middle Kingdom in the mid 2nd millennium BC, the pharaohs created an overland route from the Red Sea to the Nile Valley.

Some early cultures never mastered the art of road-building. The Minoans on Crete created a small network in the 2nd millennium BC, but the Mycenaeans, their contemporaries on the Greek mainland, made do with tracks.

Sturdy and slow
*Ox-drawn wagons like
the model above travelled Chinese
roads in the 6th century AD.*

Roman milestone
*A limestone route-marker
that once stood beside the
road linking Dijon and
Bordeaux in France, where*

THE SILK ROADS

The term 'Silk Road' was coined in the late 19th century as a poetic name for the overland trade route from China to the eastern Mediterranean coast and so on to western Europe. In fact, there was more than one silk road – a whole network of interconnected routes snaked through Central Asia and Iran. The Han Emperor Wudi (140–87 BC) was the first to encourage the development of trade with people who were several months' journey to the west. Silk was a coveted luxury all the way from Persia to Rome and could be exchanged for tea, horses and other imports. Merchandise was passed from trader to trader en route. The roads remained in use for 15 centuries, serving as channels for cultural as well

Bringing people together

The growth in trade encouraged the development of road networks, but a more important factor was the emergence of strong states with mobile armies. The rulers of the first empires knew that good communications were essential to unify the lands under their control. The Assyrians developed a highway system by the 9th century BC. In the 6th century BC Darius I of Persia linked up existing routes to create a network that became the envy of neighbouring lands. He commissioned the Royal Road that led from his capital at Susa to Ephesus in what is now Turkey, as well as another trunk road from the Mediterranean to the Persian Gulf.

From 221 BC Qin Shihuangdi, the first emperor of China, pulled an ill-assorted collection of separate kingdoms into a centralised state. The transformation involved building a highway network that eventually stretched over 7,000km (4,350 miles). He also put in place a postal relay system to carry official messages across the realm, and even standardised the axle length of carts so that all would fit the ruts in the new roads. The goal was to remove obstacles to the circulation of goods and people across his domain.

The road to Romanisation

None of these early endeavours matched the ambition of Rome's rulers, who were the first to adopt a joined-up strategy for land communications across their territories. Until about 300 BC Rome was the centre of a purely local network linking the principal towns of Latium, the coastal plain around Rome. In the following century the expanding Republic committed itself to a civil-engineering project that matched its political ambitions. A series of new roads were built, all linked to colonising ventures and each named for the magistrate who commissioned it. The most important was the Via Appia ('Road of Appius'), which became the main east–west highway linking Rome to the port of Brindisi on the Adriatic coast and so to Greece and Asia beyond.

The network continued to expand with the coming of the Roman Empire in the 1st century BC. Commissioners were appointed with specific responsibilities for the construction and maintenance of highways. The Romans had both the technological expertise and the ambition to undertake major civil engineering projects. They were even prepared to take on mountain ranges. In some

Arteries for armies
Troop movement was never far from the minds of the builders of Roman roads, like the highway at Carthage in North Africa (below). City streets were narrower, like this one at Pompeii (right). The raised passageways on each side served as pavements.

ROMAN ROADS

The basic techniques of Roman road-building were first elaborated in the 2nd century BC, when a distinction was drawn between *viae silice stratae* (paved roads topped with flagstones) and *viae glarea stratae* (unpaved thoroughfares made of tamped-down gravel). The precise details of construction varied from place to place, depending on the materials available locally, but the modus operandi remained much the same across the empire. Once surveyors had determined the exact route to be followed, workmen (usually legionaries) dug a ditch, if possible down to the bedrock. They then lined the trench's sides with stone blocks that served as kerbstones. The next step was to fill the bottom of the excavated space with rubble, sometimes topped with sand, before adding a layer of tamped gravel known as the *pavimentum*. To convert the road into a *via silice strata*, the *pavimentum* was then covered with paving tones fixed in cement. The surface was left gently convex to allow for drainage.

Roman road map
The road system of the Roman Empire was recorded in schematic maps known as itineraria. *The only existing copy of these, the* Tabula Peutingeriana, *is a reproduction made in 1264 and published during the Renaissance (below).*

places the military roads, like those in the region around Autun in France, were so steep that chariots had difficulty mounting them, bu the road-builders forged a way through, digging tunnels when other options failed.

Roman roads owed their reputation not just to their straightness and the quality of their construction but also to the amenities and way stations provided. Staging posts were set up at distances of roughly 15km (10 miles) where travellers could change horses. By the 3rd century AD schematised plans of the road called *itineraria* were available, providing information on routes and the facilities

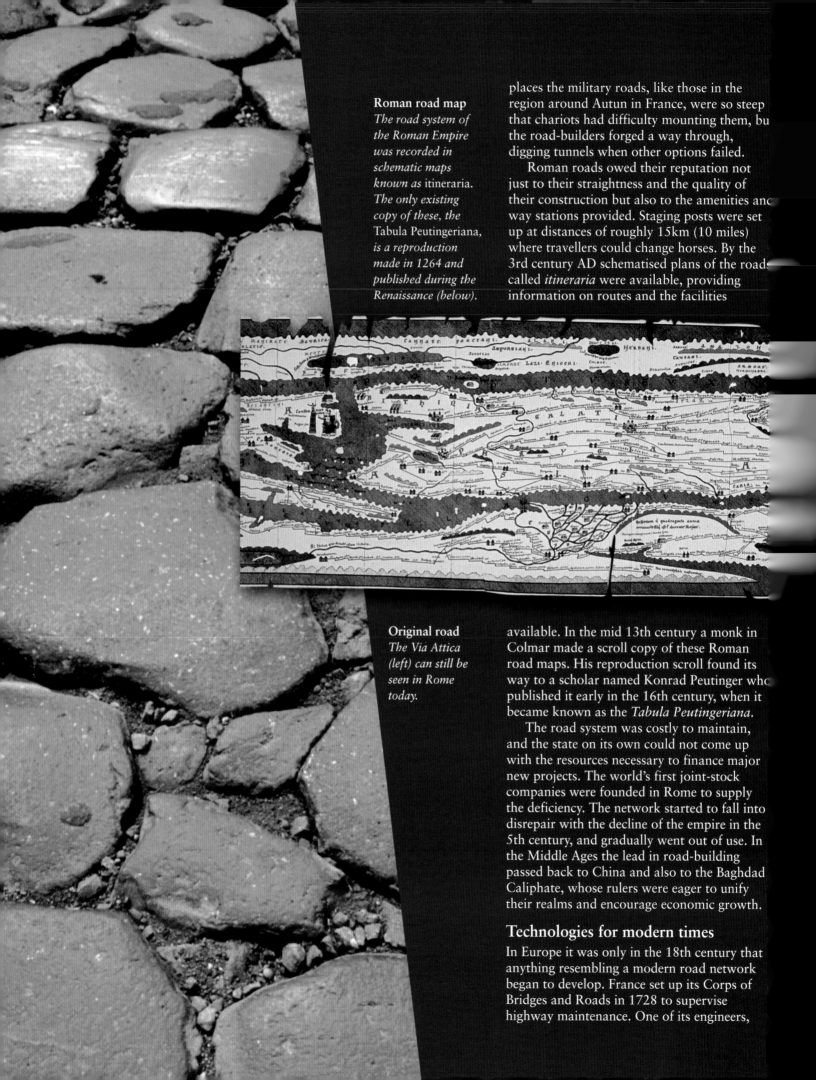

Original road
The Via Attica (left) can still be seen in Rome today.

available. In the mid 13th century a monk in Colmar made a scroll copy of these Roman road maps. His reproduction scroll found its way to a scholar named Konrad Peutinger who published it early in the 16th century, when it became known as the *Tabula Peutingeriana*.

The road system was costly to maintain, and the state on its own could not come up with the resources necessary to finance major new projects. The world's first joint-stock companies were founded in Rome to supply the deficiency. The network started to fall into disrepair with the decline of the empire in the 5th century, and gradually went out of use. In the Middle Ages the lead in road-building passed back to China and also to the Baghdad Caliphate, whose rulers were eager to unify their realms and encourage economic growth.

Technologies for modern times

In Europe it was only in the 18th century that anything resembling a modern road network began to develop. France set up its Corps of Bridges and Roads in 1728 to supervise highway maintenance. One of its engineers,

In the 15th century a road more than 5,300km (3,300 miles) long linked the Inca lands, running from Quito in what is now Ecuador to Cuzco in Peru, their capital, and on to Argentina. Hewn from rock, the route was paved for long stretches. Way stations known as *tambos* were built at intervals of about 20km (12 miles).

Remote city
Machu Picchu lay off the main Inca highway network.

Pierre Trésaguet, developed a new method of construction partly inspired by the Roman example. The roads were kept straight wherever possible and made of paving stones or a thick layer of gravel set on a stone foundation. A course of sand was sometimes added. The surface was usually crowned to allow for rainwater runoff. By the middle of the century Trésaguet's techniques had been adopted across the continent.

The next major advance was in England, where a road refurbishment programme was undertaken at the start of the 19th century. John McAdam, general surveyor for the city of Bristol, devised a new approach to save money. He recommended stripping the soil down to the bedrock before putting in place three layers of graded aggregate – fragments of crushed recycled stone – bonded with sand and water. The passage of heavy wagons or steam-rollers served to compress the gravel, making it more compact. McAdam's methods were gradually adopted both in Britain and in France, and thanks to his efforts, the two nations had the best roads in the world by the century's end.

McAdam's innovations turned out to be unsuitable for motor cars, which threw up dust and had a tendency to skid on macadamised surfaces. From about 1920 tarmac, or bitumen, came into use instead. In spite of growing competition, first from railways and later from air transport, roads continue to carry the bulk of passenger and freight traffic, and the highway network is still expanding.

Heavy traffic
Steam rollers like this giant machine began to be used to smooth newly laid road surfaces in the 19th century.

BRITAIN'S TURNPIKE SYSTEM

By the late 17th century the growing use of wheeled vehicles had made many of Britain's roads all but impassable. In response, in 1706 Parliament set up the first turnpike trusts, which were authorised to erect gates and charge tolls to travellers in return for keeping the highways in a good state of repair. The system fell into disuse with the coming of the railways. The last trusts were abolished by the Local Government Act of 1888, which passed responsibility for road maintenance to county councils.

The sum of knowledge in a single book

The goal that the creators of encyclopaedias set themselves could hardly be more ambitious: to assemble, organise and condense the entire existing body of knowledge into a single work. Tradition holds that the first such enterprise was undertaken by a Roman named Varro, whose *Antiquitates* ('Antiquities') was published in about 47 BC.

Marcus Terentius Varro was described by the rhetorician Quintilian as 'the most learned of Romans'. His *Antiquities*, dedicated in 47 BC to Julius Caesar, ran to 41 volumes, a vast compendium that was almost certainly the world's first encyclopaedia. It started off in the realm of mythology and went on to cover the sum total of knowledge in Roman culture at the time. It was to become an invaluable source for later commentators on ancient Rome, who borrowed from it at length. Sadly, only a few fragments of this great work remain. Other works by Varro that have survived include a three-volume work on agriculture and six volumes of a 25-volume set on the Latin language, as well as fragments of the *Disciplinae*, another work that sought to summarise the knowledge of the day categorised under the disciplines associated with the nine Muses.

Varro's most distinguished successor was Pliny the Elder, whose monumental *Natural History* appeared in AD 77, dedicated to the Emperor Titus. Running to 36 volumes, it covered the structure of the universe, geography, zoology, plant life, mineralogy and medicine. It was accompanied by an additional volume detailing the sources that Pliny used in preparing his work; more than 100 authors were credited.

Compilation and classification

In citing his sources Pliny made it clear that, like Varro, he was not presenting original research. Instead he was compiling information from existing works in such a way as to make it readily accessible. In short, he was fulfilling the task of an encyclopaedist.

Roman scholar
Shown here in a 15th-century Italian painting (above), Pliny the Elder was one of the first encyclopaedists.

Early Chinese reference works
The two Chinese woodblocks below show a fishing rod with reel (left) and a counterweight system for drawing up water from a well (right).

CHINESE ENCYCLOPAEDIAS – SELF-CONTAINED LIBRARIES

Chinese encyclopaedias developed in the context of the Confucian classics. Would-be government officials had to master the Confucian texts when studying for the imperial exams introduced during the Tang era (618–907 AD). Some texts were commissioned by the court, bringing together existing works on different subjects. The first attempt at a comprehensive compendium, the *Taiping Yulan*

or 'Imperial Readings of the Taiping Era', was completed in 983, citing more than 2,000 sources. Technical compilations devoted to specialist subjects first appeared in the 17th century. The world's biggest encyclopaedia, the *Gujin Tushu Jicheng* ('Complete Collection of Writings from the Earliest Times'), was completed in 1726. It ran to 800,000 pages, all the while retaining the traditional anthology format.

Pliny was well aware of his own limitations, stating in a preface 'I do not doubt that I have left out many things'. Yet despite its occasional vagaries – Pliny classed the unicorn among living creatures, for example – his work remained a principal reference source throughout the Middle Ages.

Encyclopaedias in the Islamic world

Islamic encyclopaedias hold an important place in the history of recorded knowledge, as the Arab world played a central role in preserving the heritage of Classical antiquity and transmitting it to the West in the medieval period. The first Arab compendiums saw the light of day in 9th-century Baghdad, capital of the Abbasid caliphs, following the foundation there of the House of Wisdom, a library and place of study that brought together scholars and translators of works from Greece, Persia and India. Researchers seeking to systematically arrange the information aimed to compile what a 'wise and learned man needed to know'. The defining formula came from al-Masudi, whose *Meadows of Gold*, covering history and geography, was completed in 947.

The scientific spirit

Although many Islamic encyclopaedias took the form of anthologies, it would be wrong to dismiss the Arabic contribution as simple compilations of quotations and knowledge from earlier sources. To take one prominent example, al-Biruni's 11-volume

An Arabic encyclopaedia
Illustrated entries on optics (below) and botany (right), taken from a 9th-century Arabic compendium of knowledge.

THE FIRST VERNACULAR ENCYCLOPAEDIA

As the language of scholarship in medieval times, Latin was the natural medium for encyclopaedias, even though this meant that few of the emerging middle classes, however hungry for knowledge, could read them. The situation only began to change after 1265, when the Florentine Brunetto Latini published his *Li Livres dou Tresor* in a dialect of French.

Medieval treasure
Written in the Picardy dialect of French, Li Livres dou Tresor survives in a beautifully illustrated parchment copy (below left).

Masudi Canon, prepared early in the 11th century, was the work of a true scientist whose interests stretched to mathematics, astronomy, mineralogy, geography and cosmology. The Persian philosopher Ibn Sina, better known in the West as Avicenna, was a leading light in the golden era of Arabic scholarship. He set out to gather all the medical knowledge of his day in his *Canon of Medicine*. He not only included scientific theory but also gave the work a practical structure, grouping diseases by the organs they affected and moving from general points to pathologies and then to cures. His work remained in use as standard medical reference in Europe as late as the mid-17th century, and Avicenna became known as the father of modern medicine.

The grip of religion

European encyclopaedias of the medieval period started from the premise that the Christian faith underpinned all knowledge. The *Etymologiae* of Isidore of Seville (*c*560–636) and *De Universo* ('On the Universe') by Rabanus Maurus (*c*780–856) both contained much scriptural analysis. In the mid 13th century the Dominican friar Vincent of Beauvais produced his *Speculum Maius* ('Great Mirror'), a monumental work of history, arts and sciences, along with a survey of the natural world described in the order of its creation according to the Book of Genesis.

It took the 16th-century Humanist movement in Europe to free encyclopaedias from the shackles of theology and restore them to their antique role as collections of the branches of knowledge. Kendal-born Ephraim Chambers prepared the first edition of his *Cyclopaedia*, 'An Universal Dictionary of Arts and Sciences', while apprenticed to a London globe-maker. It appeared in 1728 and was hugely influential, introducing the innovation of cross-referencing between articles. The first modern encyclopaedia as we know it, the 35-volume French *Encyclopédie*, began life as a translation of Chambers' work but developed a momentum of its own under the editorship of Denis Diderot and Jean d'Alembert. Prepared between 1751 and 1780, it is considered a masterwork of the Age of Reason.

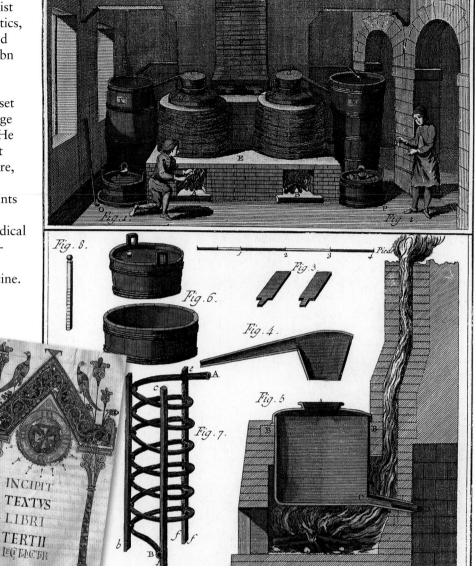

ARRANGING ENTRIES IN ALPHABETICAL ORDER

Early encyclopaedias were mostly organised on thematic lines, but the growing amount of knowledge made that arrangement hard to sustain. In the 17th century a number of works were published that listed entries alphabetically. The first work to adopt the alphabetic system was the *Grand Dictionnaire Historique* ('Great Historical Dictionary'), which appeared in France in 1674. The first English alphabetical encyclopaedia was John Harris's *Lexicum Technicum*, consisting largely of explanations of terms used in the arts and sciences, the first volume of which appeared in 1702. Britain's answer to the French *Encyclopédie* was the *Encyclopaedia Britannica*, published in Edinburgh between 1768 and 1771 and the oldest alphabetical English language encyclopaedia still in print.

Two classic works of reference
Eleven centuries separate Isodore of Seville's Etymologies *(above left), written in the 7th century, from the first truly modern reference work, the* Encyclopédie *edited by Diderot and d'Alembert in the mid 18th century (above).*

The lift *c*30 BC

Vitruvius's lift
A medieval lady lifts her husband using a pulley mechanism of the type described by Vitruvius in the 1st century BC.

It seems likely that some Roman buildings may have had lifts operated by slaves. Certainly the architectural writer Vitruvius, who lived from about 70 to 25 BC, described the principle of a platform contained within a vertical shaft that was raised and lowered with the aid of a counterweight and a pulley with a handle. Such machinery was probably used in the same way as the crane, which made its appearance at about the same time, to lift heavy weights such as blocks of masonry in construction. Louis XV of France is said to have had a 'flying chair' installed in the Palace of Versailles in the 17th century that operated very much along Vitruvian lines. The same idea resurfaced in the dumb-waiter arrangements that became popular after kitchens were relegated to the basements of stately homes.

Hydraulic lifts

The inventor of the flying chair was an aristocratic engineer by the name of Jean-Jacques Renouard de Villayer. His chief innovation lay in designing

THE COMING OF ELECTRICITY

The first electric elevator was unveiled in 1880 by Werner von Siemens, founder of the German engineering firm of that name. His design featured a motor attached to the bottom of the cab, which was driven up a shaft by pinions that interlinked with racks fitted to the shaft walls. An electrically driven pulley system was introduced in 1887. By the turn of the century motor technology had advanced sufficiently to allow the development of elevators serving high-rise buildings.

a self-operated system by which the users of a lift could haul themselves up with the aid of a pulley. The problem of energy sources to power such devices was to loom large in the years to come. In 1835 a couple of English engineers named Frost and Strutt introduced the 'Teagle', the world's first steam-powered, belt-driven lift. Fifteen years later an American inventor, Elisha Otis, developed a safety device that prevented lifts from falling. He showed it off at the New York World's Fair in 1853 by having an axeman cut the hauling cable above the platform on which Otis stood; the lift fell a few inches, then stopped.

The demonstration was so convincing that Otis's company received a commissioned to install the world's first passenger elevator in a five-storey building in New York four years later. Otis's invention, which he called the 'safety hoister', began the skyward development of American cities, but today's skyscraper cityscapes were only made possible by continuing improvements in elevator technology, many of them associated with US engineer Frank Sprague (1857–1934).

Elisha Otis's demonstration
The American inventor of the safety lift shows off his invention at the New York World's Fair in 1853 (right).

Harnessing the power of water

The principle on which a watermill works is straightforward enough: the power of falling water is converted into mechanical energy, thereby saving human and animal effort. Watermills were used initially to grind grain and for centuries they remained the most powerful energy source available – one that was entirely renewable.

Miniature model
Made in the 11th century, this water-powered Chinese clock mechanism is a perfect model in miniature of the vertical waterwheels used in mills.

Writing in the 1st century BC, the Roman architectural author Vitruvius left the first detailed description of a watermill to have come down to posterity. He described a vertical-wheel system, employing gears, that was driven by the current of a millrace (a fast-moving stream). Vitruvius called it a hydralete, a Greek term, which has led scholars to conclude that the idea originally came from Greece and was improved by the Romans.

The hydralete was far from being the first experiment with water power. Since the 4th century BC, people in China had been using hydraulic energy to operate twin-piston forge bellows. And the idea for the watermill itself may have come from the noria, a device used in the Middle East from early in the 2nd millennium BC to raise water for irrigation (see page 61). Norias took the form of large waterwheels equipped with buckets. They were turned at first by men or draught animals, but after a time someone had the idea of attaching paddles to the wheels so that they revolved simply by the force of the current.

Horizontal wheel
Mills of this type were popular in the 19th century in tropical latitudes, where they were used to crush sugar cane.

A simple yet durable system

The Romans in fact built two different types of watermill. The simplest had a horizontal wheel mounted on a vertical shaft and placed in such a way that the flow of the current struck wooden paddles attached to its rim. The shaft directly connected the wheel to the millstone. Requiring no gearing, this system was

THE MILLS OF BARBEGAL

Early in the 4th century an extraordinary milling installation was erected at a place called Barbegal, near Arles in southern France. Sixteen waterwheels were arranged in two parallel rows, fed by water from an aqueduct. French historian Jean Gimpel claimed that the Barbegal mills, which between them reportedly produced some 28 tonnes of cereals daily, were 'the largest industrial complex known to have existed in the Roman Empire'.

probably the first to come into use. It remained in operation in Scandinavia until recent times, and is still sometimes referred to as a 'Nordic mill'. Easy to use but low-yielding, it was mostly suitable for small-scale operations.

The other type of mill had a vertical wheel attached to a horizontal shaft. This arrangement only became practical once a way had been found to transmit energy through 90 degrees, for the shaft that connected to the millstone was vertical. The solution lay in a system of gears featuring a wooden disc crowned with pegs. A disc of this type was attached to the horizontal shaft and linked with a gear drum attached to the vertical shaft; the pegs engaged and the job was done. A vertical-wheel mill could deliver energy equal to three horse-power, producing a much better yield than the horizontal model, and was therefore better suited to industrial uses.

Wooden gearing

Over time the operation of mills gradually grew more efficient. The angle of the paddles on horizontal waterwheels was inclined to maximise the amount of energy transmitted from the current. In the earliest vertical-wheel mills, the current struck the base of the wheel; such mills are described today as 'undershot'. In the latter days of the Roman Empire, in the 5th century AD, engineers added a bucket arrangement to the paddles on vertical wheels and directed the millrace at the top, rather than bottom, of the wheel. As a result, the weight of the water falling into the buckets

now drove the wheel through the force of gravity, and the 'overshot' mill was born.

Whether horizontal or vertical, the wheel was usually made out of oak and the gears of elmwood. Cast-iron paddles made their first appearance in the 18th century, after which metal gradually became the norm.

Mills for all seasons

The first watermills were used to grind grain and press olives, the continuous circular motion of the wheel precisely mimicking that of the millstone. The introduction of camshafts in the 10th century provided a mechanism by which the continuous circular movement of the

VITRUVIUS – HISTORIAN OF TECHNOLOGY

Vitruvius (c70–25 BC) was an architect who also served as Julius Caesar's military engineer. Toward the end of this life he wrote his *De Architectura* ('On Architecture'), seeking to pass on all that he had learned in the course of his career. Divided into ten books, the work was encyclopaedic in scope and it became the only Classical work on architecture to survive into the modern era.

Always emphasising the importance of harmonious proportion in buildings, Vitruvius covered subjects as diverse as the design of the capitals topping columns, water treatment methods and town planning. He wrote eloquently of the proper training for an architect, maintaining that 'architecture is a science that needs to be supported by a wide variety of studies, permitting the practitioner to assess the quality of work in all the other arts associated with it. Such knowledge comes from both theory and practice.' In the tenth book he described a variety of different machines including the tympanon (a precursor of the piano), the hydraulis or water organ, and Ctesibius's water-pump. It was Vitruvius, too, who left the first account of the vertical-wheel watermill.

Vertical wheel
Driven entirely by the force of the current, provided in this case by the River Seebach in Austria, watermills are showcases of renewable energy (above).

FLOATING MILLS

In the year 537 Rome was besieged by the Ostrogoths and the aqueducts feeding the city's watermills were cut off. To maintain the supply of flour, the Byzantine general Belisarius came up with the idea of setting mills on boats, their wheels turned by the flow of the River Tiber. From that time on the concept of floating mills spread across Europe. In the 16th century 68 of the devices were in operation on the River Seine in Paris. Eventually they started to impede river traffic, and they gradually fell out of use in the course of the 19th century.

Handy hydropower
Floating mills were once common on Europe's waterways, as this 18th-century engraving of the River Seine at Paris shows.

Inner secrets revealed
The mechanism of a watermill, as illustrated in John Bate's The Mysteries of Nature and Art *(1635).*

waterwheel could be transformed into straight-line motion. In such cases the mill's driveshaft was equipped with protruding lobes – the cams. As the shaft turned, the cams pressed down alternately on the handles of tools – hammers, for example – that pivoted on another shaft. When the pressure was removed, gravity caused the hammerheads to fall. In fulling mills, for example, hammers operated by wheel rotation kneaded the cloth to be scoured. Similar pounding mechanisms were employed in hemp mills and stamping mills, which

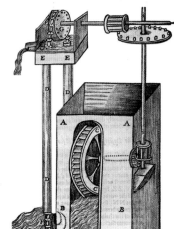

started to appear from the 11th century. By adding a spring to the camshaft arrangement mills could be made to drive saws. In time bellows in forges and in paper and gunpowder manufactories all drew their energy from waterwheels.

The big wheels could also be used to supply water. In the late 17th century a complex of 14 separate paddlewheels served the French town of Marly-le-Roi and the nearby palace of Versailles. The system was wasteful, delivering less than a quarter of the expected water output because of gearing inefficiencies. But despite its limitations hydraulic power would play a crucial role in the early Industrial Revolution. In 1771 Richard Arkwright, a pioneer of mechanisation in the cotton industry, built the world's first water-powered mill at Cromford in Derbyshire.

A late developer

The historian Marc Bloch noted that 'even though the watermill was invented in antiquity, it only really came into its own in the Middle Ages'. Its slow development can be explained partly by the fact that Roman society had cheap and plentiful labour in its slaves, and so had relatively little need for hydraulic power. Another factor is that the flow of rivers on the Italian peninsula tends to be irregular, so mills had to be linked to aqueducts, making them expensive to set up. Water mills appeared in England in the 8th century and by the time the Domesday Book was compiled in 1086 there were over 6,000, mostly in East Anglia and the Midlands, where they served the burgeoning wool industry. One significant entry was the earliest recorded tide mill, in Dover harbour.

Between the 11th and 13th centuries Europe experienced rapid population growth and the need for mechanical assistance became urgent. Cistercian monks quickly came to understand the significance of hydraulic energy, and they took care to establish their abbeys near fast-flowing streams. A 13th-century chronicler wrote of the monastery at Clairvaux: 'How many horses would exhaust themselves, how many men would wear themselves out accomplishing the work that is done for us by the gracious river to which we owe our clothes and our food! Setting so many wheels turning rapidly, the stream still spurts forth foaming; one might almost say that the water itself was ground and crushed. Next it flows into the tannery, where it prepares the leather for our brothers' shoes; showing as much energy as care, it then divides into innumerable small channels to perform a variety of functions ... from cooking and sifting to watering, washing and grinding, and never withholds its support.'

With its mostly temperate climate, Europe as a whole benefited from abundant and reliable hydraulic resources. As a result, mills were eventually erected on every suitable stream, barrages and weirs were constructed to feed millraces, and channels were dug on the slopes to harness the power

Mechanisation of labour
The Middle Ages saw greatly increased use of watermills, particularly after the Renaissance.

of mountain torrents. By the mid 19th century there were reckoned to be some 100,000 watermills in operation across the Continent.

By then, the steam engine was starting to challenge the watermill's supremacy. After a curious interval in which the new machines were employed within mills pumping water to turn the wheels, the superior energy output of steam engines was recognised and the watermills gradually fell into disuse.

Putting water to work
The Moors brought many innovations in water use to Andalucia in Spain, including this waterwheel on the Guadalquivir River.

THE ROMAN HOUSE
Luxurious villas and crowded tenements

Reflecting the social hierarchies of the day, Roman houses displayed conflicting cultural influences. They also showcased the expertise available at the time in construction techniques, interior decoration and all the arts of comfort.

Gaius Asinius Pollio had every reason to be satisfied with the vast new house going up for him on Rome's Aventine Hill. The year was 35 BC or thereabouts, and Pollio was one of the wealthiest men in the city. Much work remained to be done. Although the building itself was in place, the decoration had only just got under way. Labourers were preparing the undercoats for the mural paintings designed to enliven the walls. One man was grinding pigments – red, blue and yellow, along with black and white – in a mortar for the foundation layer. Another was mixing the coloured powders to create intermediary shades. Pollio was particularly pleased with the spaciousness of his dwelling, for sizeable building plots were much in demand by the wealthy in a city packed with apartment blocks housing the poor. Sites normally became available only after a disaster – a fire, perhaps, or the collapse of a poorly constructed tenement. But such events were hardly rare and Pollio had been able to find himself the space to build a home worthy of his station.

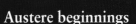

Austere beginnings

Roman homes underwent a complex evolution, from the rudimentary country villas of the Etruscan period to the luxurious dwellings of late Republican times. In the early days the *domus*, or town house, consisted of a small courtyard – the atrium – and a single room (that of the master of the house), surrounded by enclosing walls. The courtyard was part covered by an inward-sloping canopy roof pierced by a rectangular opening at the centre.

Over time other rooms were added around the sides of the courtyard and the atrium design became the model for all later developments. Initially simple and austere, the *domus* gradually became bigger and more luxurious under Greek influence from the 4th century BC on, although the exterior retained its original plainness, with blank walls, little or no decoration, and at best one or two narrow windows to provide ventilation. The heart and soul of the dwelling lay inside, providing an inner sanctum for the family. Symbolically, the atrium also sheltered the *lararium*, a shrine dedicated to the household gods believed to watch over the home.

Making an entrance
A mosaic depicting sandals (above left) marked the entrance to a Roman settlement in what is now Algeria.

Shedding light
A lampstand dating from the 1st century AD. Oil-lamps were suspended from its branches.

Ancient decor
The triclinium or dining-room of a villa excavated at Herculaneum (left) features a mythlogical wall mosaic depicting Neptune and Amphitryte.

The House of Lovers
Frescoes like this banquet scene preserved in Pompeii were a favourite wall decoration in well-to-do homes.

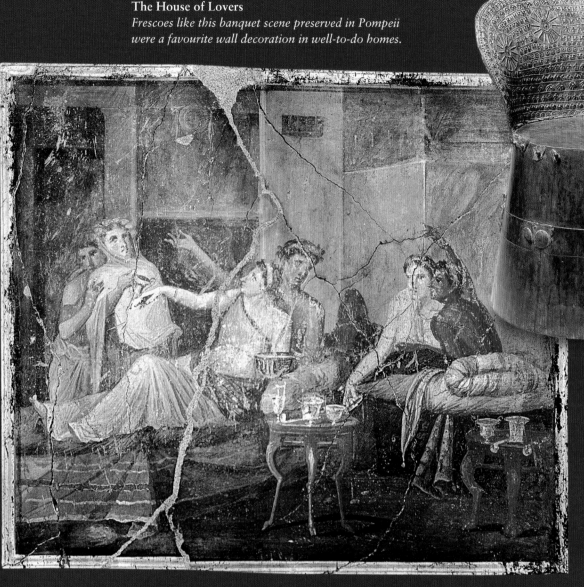

Etruscan seat
A bronze chair, decorated with geometric patterns, from the Etruscan period in Italy's history, in the 7th century BC.

The Greek influence

In the countryside, as in the town, only the wealthy could afford courtyard houses, and for the most part they had clear ideas about what they wanted and how to get it. All the villas were built to a similar general plan reflecting Greek models. A single doorway opened from the street, providing access to a central atrium open to the heavens. Here, the main life of the house was focused. The *tablinum*, an office or study for the master of the house, gave off the atrium, as did the other living rooms.

Most of the light came from the atrium's open roof. This also admitted rainwater, which was gathered in the *impluvium*, a rectangular pool connected to an underground cistern. From the 1st century on, large houses had a second open space known as the *peristylum*, a garden colonnaded on two or three sides that was often planted with rare plants and decorated with fountains.

A TASTE FOR REFINEMENT

Following Rome's conquests in the eastern Mediterranean in the 2nd century BC, the influence of Hellenistic art made itself felt in well-to-do Roman homes. One example was in the arrangement of the *triclinium* or dining-room, which came to be furnished in the Greek manner with three couches for diners. Wealthy Romans sought out rare woods to copy oriental models, encrusting them with ivory and tortoiseshell and decorating them with gold, silver and bronze. In general houses were sparsely furnished, and individual furnishings were often moved from room to room.

Interior decoration received a great deal of attention. Oil lamps and candles set in candelabras lit the rooms. Mural paintings representing mythological subjects or exotic scenes decorated the walls; sometimes these took the form of *trompe-l'oeil* windows opening onto imaginary landscapes. Floors were covered with mosaics made of tiny terracotta cubes bonded with a cement made with bitumen, gum or resin. Mosaic designs often took the form of geometric patterns, although they sometimes featured scenes similar to those on the walls.

Roman villa
A model of a villa at Laurentium, a seaside resort near Rome, based on a description by the author Pliny the Younger, who owned the house in the 1st century AD.

Building blocks

The principal construction materials were wood, brick and stone, which at Pompeii took the form of tuff, a volcanic lava. Building techniques changed over time. The oldest houses were constructed of masonry blocks bound with clay, but the invention of cement in the 1st century BC changed the way that masons worked. Thereafter, outside walls were either built of stones bound in mortar (a style known as *opus incertum*) or set at an angle to create a diamond-shaped effect (*opus reticulatum*). Corner walls often featured alternate courses of brick and stone, the so-called *opus mixtum*. The introduction of vaulting in the 4th century BC made it possible to put up higher buildings and have roofs with a wider span.

Housing the urban poor

As the capital of the empire, the city of Rome was a vibrant hub and a magnet for newcomers, who often had difficulty finding accommodation. By the 1st century BC the city housed maybe a million inhabitants, by far the biggest city of its day, and developers had to use considerable ingenuity to house them all

between the conurbation's seven hills. Every available scrap of land was put to use, and apartment buildings rose on all sides. The *domus* now became a luxury, a status symbol for the upper classes, as it took up more land than most citizens could afford.

For want of space, the city's contractors built upward. The apartment blocks known as *insulae* were put up quickly and cheaply. Usually they too were constructed around an internal courtyard that supplied indispensable light. Rents were high despite the shoddy quality of the accommodation provided. The ground floor was usually either let to a single tenant or occupied by shops, but the upper storeys were often sublet several times and were divided up by improvised partition walls or screens. For their proprietors, the *insulae* were good investments and there was no shortage of individuals eager to take advantage of the financial opportunity they presented.

Shaky foundations

Like Seneca after him, Vitruvius did not hide the fact that the Roman *insulae* were poorly constructed and had a tendency to collapse. Another danger was fire, a spectre

THE FIRST APARTMENT BLOCKS

From the 3rd century BC, multi-storey apartment blocks known as insulae were built in Rome and other cities of the empire to house the growing population. Typically comprising three floors in the early years, the blocks grew over the decades to reach seven or even eight storeys. By that time their height was cutting out the sunlight at street level, leaving lower-storey flats in semi-permanent gloom. The outer walls were most often built of brick with a stone façade, although some were constructed with a timber frame filled with a mixture of brick and stone rubble or with pozzolana and cement. It was quite common for these buildings to collapse, as they lacked adequate foundations and solid walls. Emperor Augustus (27 BC–AD 14) limited the height of the insulae to 70 Roman feet (roughly 20m). A century later Emperor Trajan reduced their height to 60 feet. At the start of the 4th century AD there were more than 46,000 such apartment blocks in Rome.

Apartment living
To house the growing population, Roman property speculators took to building apartment blocks with multiple family occupancy. This model (above) is a reconstruction of buildings at Ostia, the port of Rome, in the 2nd century BC.

Architectural variety
A fresco from Boscoreale north of Pompeii (below), suggests the wide variety of styles that could be seen in the 1st century AD.

that constantly haunted Rome's citizens. The Roman fire service dated back to the 5th century BC, but even with the help of Ctesibius's fire-pump (invented in the 3rd century BC) and the right to requisition public and private water sources, they struggled to extinguish blazes. The only other option to stop a fire from spreading was to create a firebreak by pulling down adjoining buildings that risked being set alight.

Buildings with stone foundations were relatively secure. But constrained as they were to build on a variety of surfaces, and often far from stone quarries, Roman builders increasingly preferred to use brick – and brick foundations were simply not strong enough to support the weight of the buildings erected on them, increasing the risk of collapse.

Patchwork cities

Here and there among the *insulae*, a spacious *domus* housing a single wealthy family would provide an island of light and air amid the crowded squalour. For centuries rich and poor rubbed shoulders in city streets. People lucky enough to live on the main thoroughfares had the advantage of street lighting provided by torches of pitch fixed to public buildings. The most upmarket apartment buildings even boasted lifts: wealthy proprietors would ascend to the upper-storey flats by means of a platform hauled up a shaft by slaves working with the aid of a counterweight. Most Romans had no such luxuries, even if they shared the same neighbourhoods.

City of a thousand fountains

By the 5th century AD, near the empire's end, 11 separate aqueducts brought 1.13 million cubic metres of water into the heart of Rome each day. Only the rich could afford the

Water feature
Wealthy Romans sometimes had fountains in their houses, like this one in Ostia (left)

privilege of having a direct supply into their own homes, employing spigots of a type that remained in use into the Middle Ages. Most people either got their water from one of the city's 1,352 public fountains or else bought the precious fluid from water-sellers.

Similarly, few houses had their own bathrooms. A well-equipped *domus* might have latrines installed on the ground floor. Roman attitudes to hygiene were very different to prevailing attitudes today. The toilet – consisting of a plank of wood with a hole in its centre, perched atop two uprights and set over a pit – was generally located next to, or even in, the kitchen. If the house happened to have running water, it would be flushed by waste water from the kitchen. In most cases, the pit was emptied through a specially provided sluice. The multi-occupancy *insulae* had no such amenities. Their residents had to use the public toilets, which in Rome, at least, were fairly well appointed.

Public and private hygiene

The *insulae* and many of the *domi* had at best rudimentary washing facilities, so most Romans made daily use of the public baths, which provided for all the body's needs. Up to the 1st century AD, even private houses equipped with a water-tank and drains usually had at best tiny, unheated washrooms; only luxury villas had large, well-equipped bathrooms. Elsewhere, people made do with a pitcher of water, a basin and a slop pail.

Even though the sewers built by Roman engineers were the most efficient seen to that

ROME'S WATER SUPPLY

In Rome's imperial days, responsibility for the city's water supply was entrusted to the *cura aquarum publicarum*, the imperial water board, which was directed by an important functionary, the curator of waters. He could call on the services of a corps of hydraulic engineers whose job was not just to supervise the maintenance of supply channels and water concessions, but also to construct aqueducts with the help of a labour force of some 700 slaves. The water authorities also had recourse to various private enterprises that tendered for specific jobs on the basis of specified terms and conditions.

Early tap
This bronze tap (above) came from the public Roman baths at Vaison-la-Romaine in southern France.

Roman luxury
A bath tub like this one, made of red porphyry, was the height of luxury in a Roman home.

date, most waste water still went into the street or into gutters drained by pipes running under the pavements. These were connected to the main drains, which in Rome itself meant the Cloaca Maxima. This 'Greatest Sewer' seems to have originated in Etruscan times as an open drain, but by the imperial years it had been covered over. Waste water recycled from the public baths served to flush it. The sewers emptied into rivers or the sea.

Rome's legacy

Turned inwards on its atrium and garden, the Roman *domus* combined public and private spaces within its walls without losing any of its intimacy. It offered only blank walls to the street, protecting the inhabitants from the noise and squalour outside. This arrangement became a model for urban living that was particularly popular in Mediterranean lands, where it remains common to this day.

As for the *insulae*, they too have had their successors, good and bad, down the centuries. Their modern equivalents can be seen in apartment blocks, tenements and high-rise flats in almost every city today. The quality of accommodation that they offer varies with location, price and standards of maintenance, from run-down flats to immaculate and highly desirable penthouses.

Cosmetic case
A make-up box from Cumae on the Italian coast suggests the importance that Romans put on personal appearance.

Public service
A 19th-century illustration (left) shows citizens in a frigidarium, one of the amenities on offer in Roman public baths; a slave is in attendance.

HEATING AND VENTILATION

Heating was still fairly rudimentary in Roman times. The main source was a portable charcoal-burning heater of the type known in Spain as a *brasero*. Some villas in Pompeii had an improved version known as an *authepsa*, which was also used to warm water, but this seems to have been a luxury item only available to the wealthy.

The same was true of the hypocaust, a system of underfloor heating imported from Greece in the 1st century BC by Caius Sergius Orata, a merchant from Campania. Hypocausts were mostly used in public baths, although a few well-off individuals installed them in their villas. An external furnace provided warm air that circulated in a crawlspace under the floor, which was supported on brick pillars. The heat also spread upward through the floor into cavity walls, warming the rooms without filling them with smoke.

As for ventilation, Romans had little choice as windows to the outside were few. To keep some rooms in his Golden House cool in summer, the Emperor Nero, who ruled from AD 54 to 68, chose to bury them deep below ground level in a basement. Only the houses of the well-to-do had windows of coarse glass or translucent selenite (a type of gypsum). Everyone else made do with wooden shutters or cloth coverings over the windows, so in winter they had to live with little natural light or else put up with draughts.

The art of light

Stained glass was known to the Romans, but it was only in the Middle Ages that the medium really came into its own. It emerged then to adorn the great cathedrals of Europe, bathing their interiors in glorious light. The artistic style that developed in medieval stained glass windows would have significant influence on painting and the decorative arts.

There is evidence that glass was put to decorative use as early as the 16th century BC in Egypt, when people learned to colour it by the addition of metallic oxides. But glass was prohibitively expensive to make at the time, and so was rarely used either in public buildings or private homes. Up until the Middle Ages or even later, windows were mostly closed with wooden shutters, oiled parchment or waxed cloth.

A Roman legacy

The first stained-glass windows are thought to have been made in Rome in the 1st century AD, following the discovery of a technique that involved pouring molten glass out flat on a wooden tabletop or a bed of sand. The end result gave fresh impetus to the glassmakers' art. The contemporary commentators Prudentius, Tertullian and later Paul the Silentiary all mentioned the presence of stained glass set in stucco or marble in early Christian churches, including the great cathedral of Hagia Sophia in Constantinople, although only a few fragments have survived from that time. At other sites, notably the

Basilica of San Vitale in Ravenna, craftsmen used the technique of soldering the glass pieces to one another with lead strips (called 'cames').

Medieval stained glass

Stained-glass manufacture survived the fall of the Roman Empire, but really only came into its own with the golden age of cathedral-building in Europe, which got under way in the mid 12th century. For well over a century stained glass was a principal means of expression for religious art, imposing its styles on other media such as miniature painting and frescoes. Brilliantly coloured windows reproduced scenes to edify the faithful at a time when only the educated elite was literate. The windows of Canterbury Cathedral, or those decorating the apses of the cathedral of Poitiers in France, stand witness along with many others to the vitality of an art form that was indissolubly linked to the Gothic style of architecture.

Fresh techniques and colours

Glass-making techniques barely progressed at the time. The blown glass was drawn out into a bottle shape,

Fish mosaic
This remarkably modern-looking mosaic of fish in coloured glass was actually made in Egypt in the 1st century AD.

Tinted windows
Leaded windows like this one (left) were common across Europe in the 15th century.

then split lengthwise and flattened while it was still malleable. It was then broken into coloured fragments that could be assembled with the aid of lead dividers (the cames) to form panels, sometimes reinforced by iron rods. Each light (as the separate panels were called) was then put into its appointed place in the window opening.

There was a change in the second half of the 13th century, when European artisans began to produce monochrome 'grisaille' windows for cathedrals such as Salisbury and Lincoln. These could be alternated with stained glass to allow in more natural light.

The stained-glass revival

In the 14th century a number of master glassmakers started using enamelled glass, produced by adding thin layers of different colours when the glass was blown. This technique became popular in the Renaissance years, a time when the artform of stained glass in general was in decline. Protestant iconoclasts destroyed many windows in the countries that adopted the reformed faith. Even in Catholic lands where the survival of the old faith was never in doubt, stained glass went out of fashion, for the newly popular Neoclassical style of religious architecture had little need of colour. It was only with the return to popularity of the Gothic style in the 19th century that stained glass truly experienced a revival. Among its exponents were Louis Comfort Tiffany, whose first lamps date to

1893. In the 20th century the traditional lead technique was joined by a new taste for slabs of glass set in concrete. Both approaches attracted the attention of such giants of Modernism as Matisse, Braque and Chagall.

A TASTE FOR COLOUR

Stained glass takes on its colour at the point when the silica fuses with potassium through the addition of metallic oxides. It is then reheated in an oven to anneal (toughen) it. In the 12th century copper carbonate or cobalt oxide, generally imported from Bohemia or Saxony, were used to make the colour blue. The red obtained from copper oxide was initially so dark as to be completely opaque, so instead a technique called flashing was employed by which semi-molten clear glass was dipped briefly in red to give it a thin surface coating. The semi-transparent yellow hue for which the windows of Chartres Cathedral have long been famous was achieved with the aid of silver nitrate.

The wheelbarrow c AD100

Another Chinese first
The wheelbarrow was uncomplicated in its original form, but it was to undergo various transformations over the centuries.

The wheelbarrow is a fine example of the kind of everyday object that is so much taken for granted it is easy to imagine it has always been around. In fact wheelbarrows seem to have been unknown before the 1st century AD. The earliest-known examples consisted of little more than a box supported on twin poles, with a wheel revolving on a front axle set between them. The earliest representation of one is in China, in a relief frieze on a tomb in Jiangsu province.

Powered by pushing

By employing leverage, the wheelbarrow provided a means of shifting relatively heavy weights with a minimum of effort. It was easier to use than the hods that labourers carried on their backs, or the fabric-covered panniers that had previously been in service. It was also better suited than yoked vehicles or beasts of burden for use in cluttered or confined spaces like building sites.

It is hard to say why the device, seemingly a natural extension of the wheel itself, should have taken so long to reach the West. Yet no mention of one has been found before the 13th century, when a detail from a miniature painting illustrating the French *Histoire du Saint Graal* ('Story of the Holy Grail') shows builders at work on a cathedral using a wooden barrow.

The barrows have changed little since then, except to benefit from some subsequent technological improvements. By the 18th century, two-wheeled designs were common, and favoured for their efficiency by the US

President and polymath Thomas Jefferson. Today's models are equipped with metal wheels with rubber tyres, for example, and the barrow itself is more often made of steel or galvanised iron than of wood so as to take advantage of the metal's lightness and increased resistance to wear and tear. There are now also four-wheel versions, as well as barrows that can be folded away for easy storage. The ultimate luxury comes in the form of motor-powered wheelbarrows, which are even more sparing of human energy.

Human load
Wheelbarrow-like vehicles were also used to carry people. This illustration of a bad-tempered passenger is from the 15th-century.

Bath chair
A model of a 17th-century French vinaigrette (left).

HUMAN-POWERED CARRIAGES

The wheelbarrow concept has been adapted in various times and places to create passenger-carrying vehicles, the best known being the rickshaw of East Asia. The model shown here represents a French *vinaigrette*, so called because it resembled the barrows used by vinegar-dealers. In Britain human-powered Bath chairs were used by invalids, initially in the city of Bath itself, hence the name.

Scissors *c* AD 100

Uncertain origin
The Romans are credited with inventing scissors in the 1st century AD, but this Chinese pair (below) also dates from roughly that period.

Cross-bladed scissors made of two iron cutting blades pivoting on a central axis were a Roman invention of the 1st century AD. Before that there had been so-called 'spring scissors', used from early times in Egypt and the Far East by barbers and tailors, and also for cutting up animal hides. These took the form of two blades connected by a thin strip of horseshoe-shaped metal that acted as a spring. The Egyptians later produced an improved version of the design with a detachable blade that could be replaced when it became blunt and also used on its own as a knife. Scissors of this type remained in use until the Middle Ages, showing up in miniature paintings dating from the 13th and 14th centuries. Medieval miniatures also reveal that cross-bladed scissors took their familiar form from the 10th century on, with the addition of ringed handles to make them easier to grip. Scissors were valuable items in the Middle Ages and ladies would keep their scissors in iron or embossed leather cases.

Scissor design has progressed over the centuries. Later developments include pinking shears, invented in 1893 by Louise Austin of Whatcom, Washington, USA. By cutting fabric in a zig-zag, pinking helps to prevent the cut edges from fraying.

Cutting cloth
The scissors used by drapers in the 15th century were still cumbersome (left), resembling the hand-clippers used to shear sheep.

FROM IRON TO STEEL SCISSORS

Scissors gradually spread across Europe in the Middle Ages and early modern period, but were not mass-produced until the 18th century. The pioneer of this industrialisation was Robert Hinchcliffe, who first used cast steel to manufacture scissors in Sheffield in 1761. To this day steel is preferred for making the blades, thanks to its lightness and resistance to rust; metal with a high carbon content is used for precision implements, while stainless steel is preferred for surgical instruments. Much thought has also been given to improving the design to adapt it to different uses, the shape of people's fingers and for the left-handed. Today, many pairs of scissors have plastic handles.

The evolution of modern numerals

Counting symbols
These tokens were used as counting aids in c3300 BC at Susa, in what is now south-western Persia.

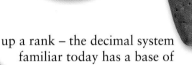

It sometimes seems hard to imagine that the ten decimal numerals that we use today – 0, 1, 2, 3 and so on up to 9 – were not the only system devised for counting, whether for everyday calculations or for the higher maths.

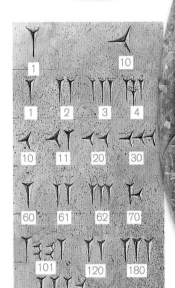

Alternative systems
Sumerian numerals were written down using nail-shaped symbols to represent units and trefoil marks to signify 10s (above). The symbols on the Mesopotamian tablet (above centre) are calculations of area made in about 2550 BC.

One bear killed, one nick made in a piece of bone, or perhaps a pebble set aside; one wolf killed, a similar mark made on a second bone, or a pebble placed on a different pile. That, or some similar scenario, may be how *Homo sapiens* first started to count. To take the word of Georges Ifrah, a historian of mathematics, a number of 35,000-year-old bones covered with notches that have been dug up at various sites in western Europe are, in effect, 'the earliest adding machines'. The system was adequate for small numbers and evidently remained in use for many thousands of years.

The need for counting

Some 5,000 years ago the Sumerians developed an arithmetical system with the number 60 as its base (the base being the number by which it is necessary to multiply a given figure to move up a rank – the decimal system familiar today has a base of 10). Mesopotamia (today's Iraq), where the Sumerian system emerged, was not only the home of the first city-states but also the scene of the first large-scale commercial transactions: stock had to be managed, accounts prepared and kept. Those responsible for keeping records had the idea of replacing the pebbles that had formerly been used as counting aids with tokens or counters whose shape and size differed according to the value conferred upon them: a small cone might represent one unit, a round token ten and a larger cone 60. The counters – representing, say, the number of goods included in a given deal – were stored in a ball of clay that was then sealed for security. To check the sum, the ball had to be cracked open.

In time the tokens were replaced by symbols in the form of holes or notches cut in a clay tablet, and in this way the first figures were born. Having already invented cuneiform writing, the Sumerians applied the same techniques to numerals. By about 2700 BC the early primitive marks had been replaced by a more sophisticated system employing nail-shaped symbols (see illustration, left).

The importance of position

Something else still had to be invented to make the counting system work properly: the positioning principle, by which the value of a given numeral changes with its place within a written number. For example, take the number 242 in our own system: the first '2' represents

2 x 100, while the second stands for 2 x 1. The numeral is the same, but the number it represents is defined by its position. The first people to introduce this innovation were the Babylonians in about the 17th century BC. By then they had mastered addition, subtraction, multiplication and division, as hundreds of clay tablets recording sums eloquently testify. Sometimes written in the Sumerian language and sometimes in Akkadian, a Semitic tongue, many were the work of schoolchildren doing arithmatic exercises; others were tables drawn up to help solve problems, such as sharing out rations or working out the interest on loans. More than 100 of the surviving tablets are concerned with geometry, providing solutions to questions involving surface area and typically concerned with fields, building sites or canals that needed digging.

The limits of addition

The Egyptians had adopted a system based on the power of 10 in the 3rd millennium BC, a natural enough choice given that people have

ten digits on their hands for counting. Egyptian maths relied on addition: each numeral had a fixed value independent of its position in the written number. To write '427', for instance, they drew four spiral shapes, two loops and seven vertical strokes. The system was cumbersome when used for very large numbers.

In about the 5th century BC the Greeks devised two separate systems. The so-called Attic arrangement combined addition and multiplication; small symbols placed inside a given numeral multiplied it by the number the symbol represented. The Ionian system used the 24 letters of the Greek alphabet plus three obsolete ones to represent the numbers 1 to 9, 10 to 90, and 100 to 900. For all its complexity, this was the system that Archimedes, Pythagoras and Diophantus used in their work to establish the foundations of modern mathematics.

The Romans, too, settled for addition as the basis of their notation from the 4th century BC. The system was based on 2s and 5s. A vertical stroke 'I' stood for the number 1, while a 'V' represented 5. 'X', meaning 10, was two Vs joined at the centre-point. The system was cumbersome and time-consuming, and was mostly used to keep records of numbers rather than to do sums. For practical calculations individuals preferred to use an abacus. Originally, these took the form of loose pebbles that could be moved around a counting board to add or multiply numbers. The English terms 'calculus' and 'calculate' can both be traced back to the Latin *calculus*, meaning a counter used for reckoning.

Mathematical table
A 13th-century-BC bas-relief from the days of the New Kingdom shows that the ancient Egyptians had a sophisticated grasp of basic mathematics.

Counting machine
A Chinese abacus from the 1st century AD (left). The abacus was an early ancestor of the modern calculator.

THE ART OF THE ABACUS

Counting on one's fingers is all well and good, but counting with an abacus works better. In the days before place value entered general usage, Greeks and Romans used the abacus for everyday calculations. The earliest mathematical aids were simple counting boards on which people made marks with a stylus. In time these were replaced by wooden frames containing parallel grooves, each representing units – ones, fives, tens, hundreds. Sums were worked out with the aid of small pebbles that the Romans called *calculi*. The familiar abacus with counters arranged on rods (left) was developed in Asia in about 300 BC. Such tools are still used alongside electronic calculators in China, Japan and Russia to this day.

The origin of zero
The idea of zero probably started out as a philosophical concept, perhaps linked to the notion of infinity. Some such thought process may be indicated by the hand gesture of the Thai Buddha (below) and also in the ring held by the Aztec god Tezcatlipoca, the Lord of the Smoking Mirror (centre). The ouroboros or serpent with its tail in its mouth (far right) was an ancient symbol of eternity.

India's contribution

Indian savants initially used an oral system under which each digit and each power of 10 had its own name. Values changed with position, so it was possible to conceive of large numbers, although each one had a separate written symbol, making complex calculations impractical. In the 3rd or 2nd century BC, a new arrangement employing the Brahmi script limited the number of symbols to nine. An historic breakthrough occurred when mathematicians brought together elements of both systems by combining the place-value concept with the nine numbers – and then added a totally new feature in the concept of zero.

Initially the zero simply served to indicate that a digit was missing. Previously the only way to distinguish between, say, 47 and 407 was to leave a space between the '4' and the '7'. Unless the writing was very neat, it was easy to miss the gap. The introduction of the '0' made errors less likely. In time, zero began to be regarded as a number in its own right. The 7th-century sage Brahmagupta defined it as the result obtained by subtracting a given number from itself.

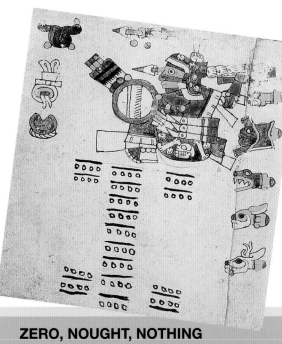

ZERO, NOUGHT, NOTHING

The Babylonians and the Mayans both had a concept of zero, representing it respectively by a double V shape set at an angle and by a snail-shell symbol used to indicate the absence of a positive number. But their noughts were not figures that could be used in calculations. The Indians who invented the modern notion of zero called it *sunya*, meaning 'void'. Arab mathematicians used the term *sifr*, 'empty', and that was how it became known to the first Europeans to encounter the figure '0'. In the 13th century Fibonacci, writing in Latin, adapted *sifr* into *zephirum*, which was later Italianised as *zephiro*. This in turn was contracted to provide our word 'zero'.

There is some uncertainty as to when the Indian breakthrough was first made, but it seems to have predated the 5th century AD. A cosmological treatise called the *Lokavibhaga* ('The Parts of the Universe'), dating from the year 458, contains the first known mention of zero combined with a place-value decimal system employing written numerals. In 499 the mathematician Aryabhata gave a full description of the system in the *Aryabhatiya*.

Arabic numerals

In the 8th century the Indian system was adopted by the Arabs, who then ruled an empire stretching from the northern borders of India to Spain. Baghdad, capital of the ruling Abbasid Dynasty, was at the time a centre of learning where scientific works from all parts of the known world were being translated, Indian mathematical treatises among them. An influential work by the Persian scholar Muhammad al-Khwarizmi entitled *The Book of Addition and Subtraction according to the Hindu Calculation* did much to spread knowledge of the Indian method. The word 'algorithm', a step-by-step procedure for solving a mathematical problem, derives from the Latinised form of al-Khwarizmi's name.

In the late 10th century the French scholar Gerbert d'Aurillac, who would later become Pope Sylvester II, spent three years studying in Spain. He returned convinced of the superiority of Arabic numerals (as the Indian numbers in their Arab form would come to be known) over Roman ones, and he set about spreading knowledge of them through Christian Europe. At first he met resistance. People used to working with abacuses showed little desire to give up their old ways of doing sums. Gerbert's task was made all the more difficult by the fact that Europe was going through hard times, with trade in decline, reducing the need for arithmetic. It was only in the

12th century that the Indian system really started to catch on. By the early 13th century Leonardo of Pisa, better known as Fibonacci, employed the decimal system with the figure zero in a treatise ironically entitled the *Liber Abaci* ('Book of the Abacus'). Even then the contest between the algorists (as enthusiasts for the decimal system were known) and the abacus-users was far from over. The abacus still has great value today as an aid to teaching simple mathematics to blind children.

Arithmetical allegory
A 16th-century woodcut shows one man employing algebra alongside another using an abacus.

Chess – c AD600

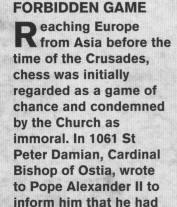

Egyptian pawn
This kneeling figure representing a bound prisoner was used in an Egyptian board game dating back to the 13th century BC.

FORBIDDEN GAME

Reaching Europe from Asia before the time of the Crusades, chess was initially regarded as a game of chance and condemned by the Church as immoral. In 1061 St Peter Damian, Cardinal Bishop of Ostia, wrote to Pope Alexander II to inform him that he had imposed a penance on another bishop whom he had found playing the game, which was already popular in the Byzantine lands.

The first known reference to chess comes in a Persian manuscript of the 7th century, which specifies that the game was invented in India in the preceding century. Historians today generally agree that it first appeared in the Indus Valley around that time. Legend has it that a Brahman named Sissa came up with the idea as an educational tool for the prince he was tutoring. Taking his inspiration from the game of Chaturanga – literally 'the four army corps'– which was played with dice, Sissa chose to represent the world in terms of a board marked off into 64 squares, over which moved different pieces representing the existing social order. The game set two rival armies against one another, the aim being to capture the enemy king. The piece representing the king was surrounded by others symbolising horsemen, chariots (now castles or rooks), and some whose significance has since been lost.

The game spread rapidly, first to Persia and then among the Arabs, who carried it around the Mediterranean Basin. It reached Europe through Spain. The name comes from the Arabic *cheik*, meaning 'lord' or 'sage'. The term 'checkmate' is said to echo another Arabic phrase, *al shah mat*, literally 'the king is dead'.

INDIAN OR CHINESE?

Joseph Needham, a historian of Chinese science and technology, pointed out that a board game featuring bronze pieces bearing markings that identified their different roles was played in China as early as the 4th century BC. This 'image-chess' was evidently linked with divination. According to Needham, the game subsequently reached India, where it influenced the development of Chaturanga. If so, chess's origins should perhaps be traced to China.

Europe's first major contribution came with the introduction of the visually friendly chequered board in about 1000 AD. Around 1500 the game changed greatly when, among other changes, the queen was transformed into the strongest piece on the board. One result was a halving of the time it took to play an average game, greatly enhancing its accessibility.

The rules took definitive form in the mid 17th century, giving each player eight pawns and eight other pieces (1 king, 1 queen, 2 rooks, 2 bishops, 2 knights), each with their own prescribed moves. The first international chess tournament was held in London in 1851, but it was 1924 before the World Chess Federation was founded. That organisation was responsible for giving the game its first universally recognised written rules.

Hebridean knight
Carved from walrus ivory, this knight is part a famous find of medieval chess pieces uncovered on the Isle of Lewis in the Outer Hebrides.

Game of kings
A manuscript illumination from the Book of Games, *commissioned by Alfonso X of Castile in the 13th century, shows two Arabs playing a game of chess.*

The pen
c AD 636

The first mention of a goose feather used as a writing implement comes from the work of Isidore of Seville, a Spanish theologian who died in AD 636. At that time parchment was gradually taking over from coarse papyrus as the favoured writing surface, and its use may have encouraged the switchover to quills, which were more supple, sturdy and less scratchy than the reed pens used previously. Medieval monastic copyists made good use of the quills' properties to show off their strokes. The plumage of other birds besides geese was also called into service; swan and guineafowl feathers were both occasionally used.

Metal pens made first of copper and then of steel put in an appearance in the 18th century. From 1818 a gilding process was introduced to help prevent corrosion. Iridium-tipped nibs were introduced in 1843, along with the practice of grooving the undersides to improve the flow of ink. Demand for pens grew as their use spread through schools and government offices. The *London Journal* confirmed their popularity: 'On the 31st of March, 1870, 2,164,320 steel pens and 553,797 goose quills were in use in the public services.'

It took half a century for metal-nibbed pens to displace quills. The new implements required a fresh style of writing. So-called 'copperplate' script was developed in England, featuring oval letters and long upstrokes and downstrokes. The next major step forward was doing away with the need for cumbersome inkwells. The American manufacturer Lewis Waterman introduced the capillary-feed fountain pen, complete with its own self-contained ink reservoir, in 1884. Disposable plastic cartridges came in with the 1950s and the modern fountain pen was born.

Rediscovering a lost art
In the last years of the 20th century, a group of calligraphers used goose-feather quills (above) to produce an illuminated edition of the Biblical gospels, reviving techniques that were current before the invention of printing.

Dancing with ink
An advertisement from the 1920s sells the virtues of fountain pens: smooth movement and the freedom associated with having an inbuilt reservoir of ink.

"Ideal" Waterman

A SERIOUS COMPETITOR

Fountain pens still have their enthusiasts today, but they now face stiff competition from ballpoints. The first reliable ballpoint pen was introduced in the late 1930s by a Hungarian newspaper editor, László Bíró, who had the idea of using printer's ink, which dried rapidly without smudging. Quickly finding that it was too viscous to flow from a fountain-pen reservoir, he tried various alternatives before discovering that it worked if a tiny metal ball replaced the nib. One of his associates, Marcel Bich, refined the process in the 1950s to launch the transparent Bic Cristal disposable pen, which remains one of the world's biggest-selling brands.

Xylography – the origins of printing

By the 8th century conditions were in place for the Chinese to invent printing – more precisely, the woodblock technique. Xylography, as it is more properly known, marked an important step in the diffusion of knowledge, for book production increased rapidly in the wake of its introduction.

Old meets new
An engraving tool of the 2nd century (top right) rests on a woodcut made in Tibet in the late 19th century.

Early subject
A printing block made in Central Asia in the 8th or 9th century shows the Buddha in the lotus position.

The preliminary steps leading to the invention of woodblock printing were probably taken in the calm of Buddhist monasteries. Reproducing images of the Buddha, along with sacred texts and magical formulae, was seen as a meritorious action for those seeking to escape the eternal cycle of rebirth. Shut off from the outside world, the monks engraved words and pictures in stone and used inked seals to leave impressions on silk, paper or plastered walls. In time wooden stamps came to replace the stone seals. In their turn, the wooden stamps gave way to blocks engraved not just with isolated characters but with entire texts carved in mirror writing, and the woodblock technique was born.

The first texts to be produced in this way were Buddhist scriptures that were printed in China. Some have been found in Korea and Japan, others in China itself in the Dunhuang Caves. They date from between AD 704 and 770.

Why in China?

The impetus behind woodblock printing indisputably came from the demand for religious texts. Yet the final step only became possible when various separate developments that had long been maturing independently in China all came together. First there was the seal, cut from jade, ivory or soapstone and used to authenticate documents. Similar devices were used in Mesopotamia, where writing was born, and in Egypt. There, however, they were cylindrical or oval in form, whereas in China they were square or rectangular, shapes better adapted to the evolution of woodblocks.

Another important element in the development of printing was the practice of stamping designs on cloth or paper, practised from the 6th century in China with the aid of carved stone matrixes. Here again, the Chinese had been using ink obtained from lampblack

THE FIRST PRINTED BOOKS

In 1907 the British archaeologist Marc Aurel Stein discovered the earliest printed book known to that date at Dunhuang in northern China. It was a copy of the Buddhist *Diamond Sutra*, printed with blocks measuring 75cm by 30cm on a paper scroll about 5m long and dated to the year 868. An even older printed text was found at the temple of Bulguksa in South Korea in 1966. It was a woodblock print of the *Dharani Sutra*, produced between AD 704 and 751.

as early as the 14th century BC, and had employed paper as the principal writing medium since the 2nd century AD.

So all the elements were in place when the spread of Buddhism gave a new urgency to the task of finding a way of duplicating sacred texts. Copying out ideograms by hand was complex and time-consuming, and led to many errors of orthography. Lacking an alphabet, the Chinese instead invented printing.

The spread of printed matter

Woodblock printing spread quickly across China. In the northern capital of Kaifeng, the nine Confucian classics were printed between 932 and 952 on the orders of a government minister named Feng Dao, having earlier been carved by imperial fiat onto stone steles. Soon private presses were turning out almanacs and herbals, works on history and books on magic. The rapid expansion of printed texts was helped by the multiplication of centres of learning under the Song Dynasty, which came to power in 960, introducing a long period of peace and prosperity. By the year 1064 the imperial library already had 80,000 volumes.

By that time woodblock printing had also established itself in neighbouring lands, notably in Korea and Japan. There, as in China itself, xylography remained the principal way of reproducing text and pictures until the 19th century, when printing presses using moveable type arrived from the West.

An enduring legacy
Woodblock printing continues to be practised in China to the present day.

THE INVENTION OF MOVEABLE TYPE

In the wake of woodblock printing a certain Bi Sheng invented moveable type at the start of the 11th century, using terracotta characters set in an iron frame. At the time, the pieces proved too fragile for heavy-duty use. Three centuries later a government official named Wang Zhen substituted wooden pieces, while also speeding up the actual setting of the type by storing the characters in a revolving cabinet, in which rhyming homonyms were grouped together. Metal type was first produced in Korea in the 15th century. Even then, the sheer number of separate ideograms making up the Chinese language ensured that moveable type never seriously challenged woodblock printing until the arrival of mechanical presses.

Keeping tradition alive
A woodblock printer practising his craft in modern Shanghai.

Musical notation – *c* AD800

No one individual could ever have claimed to have invented music. Its origins may lie in an unusually modulated word or cry, possibly just two or three sounds harmoniously combined. If people liked what they heard, they would have repeated it; then others would have taken up the refrain. But how could music be transmitted if not orally? The answer was to write it down.

Some archaeologists believe that five cruciform signs on Sumerian clay tablets dating from the 9th century BC were musical notes, but that view remains controversial. Little too is known of the ancient Greeks' way of writing down music, using letters to represent sounds with intervals indicated by special symbols. Pythagoras opened the way for this system in the 6th century BC by proposing a mathematical foundation for harmony based on the relationship between the size of a musical instrument and the sounds it produced.

Call for uniformity

The fact remains that for many centuries music was mostly transmitted orally by simple repetition. The need to write it down really only made itself felt in the West when Pope Gregory the Great undertook a major reform programme in the 6th century AD, with the aim of unifying liturgical practices throughout the Christian Church. Gregory is often mistakenly credited with inventing the Gregorian chant, which in fact developed after his day, but he did encourage the creation of a single style of Church music. To achieve that end, it was necessary to have some form of written notation.

Playing by ear
A 6th-century-BC lute player, recalled in this figure from Boeotia in Greece, would have played from memory, as all musicians did in early times.

A Carolingian invention

Gregory and his successors were concerned above all with Church unity. To achieve it they wanted worshippers throughout Christendom not only, as it were, to sing from the same hymnsheet but also in the same key – hence the need for a commonly accepted way of writing music down. By the late 8th century parchments containing liturgical lyrics were being annotated with small inflective marks called neumes that served as reminders of the general direction of the melody, indicating changes in the pitch and duration of the words to be sung rather

Neumes – the first musical notations
Reminiscent of shorthand symbols, the marks made in the margin of this manuscript of liturgical music for Easter (above) are actually neumes giving directions to the singers. The document dates from AD 860.

Musical rose
A 16th-century manuscript in circular form (right). It specifies the key, main theme and development of a canon or round. The background images shows some sample notation from a score by the Venetian master Antonio Vivaldi, written in the 17th century.

Guidonian hand
Thought to have been used by Guido d'Arezzo, the pioneer of modern musical notation, the hand served as a memory aid indicating the relationship between different notes.

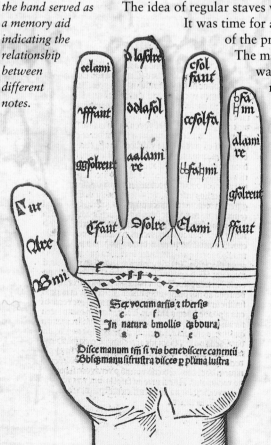

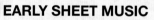

than specific notes. So a comma signified that the voice should go up, a dot that it should go down, an inverted 'v' a rapid alteration up and down.

The neume system was never intended to be more than a memory aid for people who already knew the tune, so the trend of development was toward greater complexity. By the end of the 9th century some writers were using letters of the alphabet to indicate the pitch that the neume represented. The idea of separate notes each corresponding to a sound was already appearing in embryo.

By the early 10th century neumes were being written in such a way as to indicate the outline of the melody, placed up or down on the page to suggest higher or lower pitch. Some unknown innovator now took an important step forward by adding horizontal lines to clarify the relative level of the neumes. The idea of regular staves was not far off.

It was time for a coherent synthesis of the progress made so far. The man who produced it was an Italian monk named Guido d'Arezzo, who became the principal champion of the modern stave notation, employing first four and later six lines corresponding to the six strings of the lyre. Guido, who lived from 975 to 1040, was also responsible for giving the notes the names they mostly retain to this day. He replaced the letters with the first syllables of stressed words from a Latin hymn addressed to St John, which were *ut, re, mi, fa, sol, la*. A seventh note *si*, later *ti*, was added soon after.

Continuing progress

Many other advances were made before the 17th century, when our present system of musical notation took its final form. In the 1200s, for instance, scribes using goose quills gave written notes the lozenge form familiar to this day. Minims and semibreves were added in the 15th century. In the 1600s Guido's *ut* was changed to *do* on the grounds that *ut* was difficult to sing. One only has to cast a glance at the score of a piece of contemporary music to see how complex but also how precise modern musical notation has become. Guido would no doubt have been amazed.

EARLY SHEET MUSIC

Initially monastic scribes copied music by hand in parchment manuscripts, but not long after Gutenberg introduced printing to Europe in the mid 15th century the new medium was also being used to reproduce musical scores. The first generation of printers had difficulty in transcribing the notes, which had to be copied manually into the Mainz Psalter of 1457, the earliest text to incorporate music. By the 1470s specialist craftsmen had begun to master the difficult skill of mechanically aligning notes and staves, but the process remained complex and expensive, requiring pages to be passed more than once through the press. The first single-impression scores were produced in London in about 1520.

Siege warfare
An Assyrian army led by Ashurnarsipal II uses a fortified battering ram to assault a city in the 9th century BC.

MILITARY TECHNOLOGY
Inventions born of war

Long before the invention of firearms, which would eventually revolutionise the art of war, human ingenuity had already come up with imaginative ways of spreading death and destruction.

Siege weapon
Initially developed by the Greeks, the ballista was used for firing heavy darts or stones.

In antiquity armed conflict usually meant either hand-to-hand fighting or sieges. To penetrate the fortifications of a town, the besieging forces used battering rams to break through the gates and mobile towers and scaling ladders to mount the walls. Towers remained in use for centuries. Over time they were joined by other devices designed to weaken enemy defences, including a variety of ballistic weapons invented by engineers of the Hellenistic world from the 4th century BC on.

Firing from a distance

Tradition holds that the first ruler to employ a catapult was Dionysius I (*c*430–367 BC), the tyrant of Syracuse. The implement consisted of a wooden throwing arm with a spoon-shaped receptacle at one end, drawn back by a twisted rope and pulleys. When the trigger was released, the arm shot upward until it hit a buffer, releasing the missile held in the receptacle – usually a blazing projectile or a huge stone. The ballista was a slightly smaller weapon resembling a giant crossbow; the bowstring was winched back with the aid of handles. Alexander the Great employed the ballista in his armoury, both as a siege weapon and on the field of battle.

The art of siegecraft

Demetrius I Poliorcetes, king of Macedon in the late 3rd century BC, earned his sobriquet from the Greek term *poliorketikos*, meaning

MINING AND COUNTERMINING

Besieging armies sought to undermine the fortifications of the town they were attacking by tunnelling under the foundations and then setting fires in the tunnels they had created. One way in which defenders could strike back was by digging countermines – subterranean galleries designed to break into the enemy's tunnel system so that battle could be joined underground. The Romans, who were the first to dig defensive ditches to protect their camps, also pioneered circumvallation – surrounding the enemy position with a circle of trenches, thereby cutting them off from all contact with the outside world. This technique was used successfully against the Gauls at Alesia in 52 BC.

Battering ram
This model of a Roman warship of the 4th century BC clearly shows the rostrum *– literally, the 'beak' – on the ship's bow, which was used to ram enemy vessels.*

NAVAL WARFARE

The Greeks built triremes and other light warships equipped with rams designed to hole enemy vessels below the waterline. But ramming was at best an uncertain procedure that required agile manoeuvring to be effective, and there was always a risk that the attacking vessel would be damaged as badly as its target. By the Hellenistic period the preferred tactic was boarding, often preceded by the launch of missiles. The use of ballistae and catapults at sea required the introduction of heavier vessels with four or five banks of oars. The Romans made boarding easier by equipping ships with a *corvus* or boarding bridge that hooked onto the enemy vessel.

'siegecraft'. His engineers devised a variety of war machines including giant catapults and a wheeled siege tower that was nicknamed *Helepolis*, the 'Taker of cities'. Similar wheeled towers featured in the Roman arsenal from the 1st century BC; archers sheltered behind the wooden fortifications of a tower in order to fire on defenders on the city walls. The *testudo* or 'tortoise' had a similar defensive function, in this case offering a giant shield to sappers working beneath to undermine the walls of a stronghold. Both were precursors of modern armour plating.

Medieval innovations

Such engines of war were essential adjuncts of any mighty army. At the siege of Jerusalem in AD 70, the Emperor Titus counted on the services of 300 ballistae and 40 catapults. The Romans used these devices not just in sieges but also defensively and in pitched battles. Ballistae and catapults were used into the Middle Ages; by then counterweighted devices like the trebuchet and the mangonel had been invented, with a firing range of some 200m. Such distances would hardly have impressed military engineers of the Classical world: Agesistratus, a Greek of the 1st century BC, is said to have devised a machine that threw a projectile a full 1,300m.

The first chemical weapons

Armies also had recourse to incendiary devices and sometimes to more surprising weapons. The Chinese introduced chemical warfare as early as the 4th century BC in the form of bombs filled with substances like lead oxides that gave off toxic fumes when they exploded. Another strategy involved tunnelling under enemy positions and releasing poisonous gases. Armies competed to obtain comprehensive arsenals of weapons of destruction that could be adapted to all military objectives.

Protective shell
The testudo *or 'tortoise' provided protection for Roman soldiers trying to ram or dig their way through enemy fortifications. It was the ancestor of modern armoured vehicles.*

The birth of explosives

Gunpowder was once considered the Devil's invention, but armies have been using it on battlefields for more than a millennium. The explosive powder led to the development of guns and cannons, and ultimately created a whole new kind of warfare dominated by firepower and the ability to strike from a distance.

Creating a bang
The Chinese used gunpowder to power the first guns (right), and still employ it in the manufacture of firecrackers that are let off in New Year celebrations (below).

In the German town of Freiburg-im-Breisgau stands a statue of a monk named Berthold Schwartz, long honoured by his fellow-citizens as the inventor of gunpowder and firearms in the mid 14th century. A passionate alchemist, he pursued his studies even though the Church prohibited experiments with explosives, which were considered the Devil's work.

In fact, Schwartz was far from the first person to discover the lethal powder. That distinction would also be claimed for other Europeans, among them Roger Bacon and Albertus Magnus, while a man writing under the name of Marcus Graecus (Marcus the Greek) wrote down a recipe for gunpowder at least half a century before the date claimed for Berthold. It is now known that the true inventor lived in China, where an incendiary powder based on saltpetre (potassium nitrate) was being produced as early as the 9th or 10th century AD. Thereafter the invention probably moved along the trade routes, reaching the West via Arab merchants.

子母百彈銃

Precious saltpetre

The history of gunpowder, which got its name only after the invention of cannons in the 14th century, offers an example of the way in which technologies can be progressively improved as they spread to new lands. The great contribution of the Chinese alchemists was to discover the remarkable properties of saltpetre, as potassium nitrate, a naturally occurring mineral, is commonly known. Mixed with sulphur and charcoal, it produced a powder that, while slow to ignite, had remarkable explosive properties. Initially the

CHASING DEMONS

The Chinese did not limit gunpowder to military use. They also produced paper-wrapped bangers that exploded when thrown onto fires, supposedly chasing away demons. Other fireworks burst in an array of colours thanks to the use of oxides; these, too, were thought to have power to exorcise spirits.

BAMBOO STEMS – THE FIRST FIRECRACKERS

In the 2nd century BC, long before the invention of gunpowder, people in China were already creating small explosions by throwing pieces of bamboo onto fires. The heat of the flames caused the air inside the hollow tubes to expand quicker than the bamboo could catch alight, causing the stems to explode. These early firecrackers were a popular feature of festivals and public holidays, believed to scare away demons.

them naval supremacy until the 13th century. Greek fire was a lethal mixture thought to have contained naphtha (crude oil), tar, sulphur, resin and saltpetre from naturally occurring deposits. Used to defend ports and in naval combat, it proved so effective that the details of its make-up were a closely guarded state secret. Constantine VII Porphyrogenitus, who ruled Byzantium from 913 to 959, instructed his son: 'Above all you must devote due attention to the liquid fire that is cast from tubes; and if anyone dares [ask its composition], you must rebuff the request by stating simply that an angel revealed its manufacture to the founding father of our empire, Constantine the Great.'

powder was used to produce fireworks for Daoist rituals, and the exact quantity of each of the ingredients employed was a matter of trial and error. But before long it was put to military use: the earliest known recipe for the composition of the powder appears in a Chinese military text of 1044.

In an age when towns and their fortifications were mostly built of wood, fire was a powerful weapon. Military engineers in various nations set their minds to devising prototype flamethrowers belching balls of fire, siege engines capable of hurling flaming missiles and even incendiary hand grenades, most of them unreliable and not very effective. The powder became more of a threat once armourers had worked out the most efficient formula for its composition. Soon they were placing it in hollow cylinders and using it to fire rockets, or making bombs that could be hurled by catapults, some of which released poisonous gases when they exploded. It would not be long before the first firearms appeared.

The secrets of Greek fire

Learning through traders of the special properties of potassium nitrate, the Byzantine Greeks came up with an invention that gave

Even so, the Arabs somehow managed to penetrate the veil of secrecy surrounding Greek fire. Subsequently they used it as a weapon on land as well as at sea. The French chronicler Jean de Joinville was with an army led by Louis IX of France on the Seventh Crusade and subsequently described a bombardment by the Saracens. The fire, he wrote, 'came at us as big as a barrel, with a long tail of flame stretching out behind it.'

Gunpowder reaches the West

The first Chinese experiments with using the powder to fire missiles dated from the first half of the 12th century. By the end of the century, at the time of the Mongol invasion, they had made sufficient progress to be able to shoot self-propelled projectiles from tubes.

At about the same time gunpowder became known in the West, perhaps invented independently if not brought by the Arabs. The best evidence lies in Marcus Graecus's 13th-century recipe, which specified one part of sulphur to two of carbon and six of saltpetre, and also suggested a way of purifying natural saltpetre by first washing and filtering and then crystallising it. The Arabs, meanwhile, had increased gunpowder's power

Greek fire
Even before the Chinese invented gunpowder, Byzantine fleets were using an inflammable mixture based on naphtha to create a terrifying incendiary weapon. Used in naval battles, it sometimes set the sea's surface alight.

THE MONK WHO INVENTED FIREARMS

Berthold Schwartz, also known as Berthold the Black, is a semi-legendary figure in Germany, where he is credited with the invention of firearms. Yet researchers have failed to establish when he lived or what exactly he achieved, partly because the records of the monastery in Freiburg-im-Breisgau where he worked were destroyed. Apparently he was an alchemist with a fascination for explosives as well as a Franciscan monk – an odd combination for a man of the cloth.

Early mortar
This ornate, upward-pointing cannon, dating from the 16th century, was probably less effective as a weapon than the illustration implies.

tenfold by finding a way to purify saltpetre. A 13th-century work by a Syrian engineer called Hasan al-Rammah described the process, which involved using potassium carbonate in the form of wood ash to remove calcium and magnesium salts from the potassium nitrate. Arab chemists could thus produce a purer powder that opened the way to the manufacture of explosives powerful enough to fire projectiles at high speed.

The end of a world

These technological advances enabled the makers of arms to replace the catapult and ballista of the ancient world with modern firearms, whose arrival in the West was to revolutionise the art of war. Metal cannons were first used in the 14th century, possibly at the siege of Berwick in 1319 – they appeared almost simultaneously in England, France, Italy and the Netherlands.

The earliest cannons took the form of bombards, made by forging together iron bars held in place by iron rings and mounted on a two-wheeled carriage. The first models fired arrows, but later versions hurled stone or iron balls. Bombards played an important part in the English victory over the French at Crécy in the Hundred Years' War. Thereafter cannons were standard equipment not only in siege warfare but also on the battlefield, where they could be used to harry the enemy from a distance and relieve hard-pressed cavalry and infantry units. With the coming of artillery the medieval mounted knight was made redundant and castle walls were suddenly vulnerable to cannonballs.

The introduction of 'corned' or granulated gunpowder (see box, right) in about 1420 made cannons easier to load and fire, and helped to confirm the role of the artillery as an essential arm of modern warfare. At the same time the first portable firearms began to appear, designed at first like light cannons.

History's most notorious user of gunpowder was Guy Fawkes, an expert in military explosives who, with his fellow conspirators, planned to blow up the English Parliament and kill the king, James I, on November 5, 1605. The plot failed but is still recalled each year with bonfires and fireworks.

CREATING BIGGER BANGS

Over the centuries techniques for manufacturing gunpowder changed much more than the ingredients that went into it. In the early days sulphur and carbon (in the form of charcoal) were crushed separately to form a powder. The saltpetre was mixed with water to reduce the risk of explosion and only then added to the other ingredients. The resultant lack of homogeneity in the granules slowed down combustion and limited the explosive power of the powder.

The introduction of corning – compressing and sorting the granules – addressed these shortcomings to produce a more reliable, standardised product. Yet even the new corned gunpowder sometimes combusted unexpectedly, causing serious accidents. The powder was also easily affected by humidity. As a result, the French chemist Antoine Lavoisier made further improvements to the manufacturing process when he was appointed a commissioner of France's Royal Gunpowder and Saltpetre Administration in 1775.

In the following century research into explosives accelerated: guncotton, more powerful than gunpowder but more unstable, was invented in 1846, and in 1884 a French chemist named Paul Vieille devised the first smokeless powder, a major advantage on the battlefield. Meanwhile Alfred Nobel had been working to find a way to render nitroglycerin less volatile, resulting in dynamite in 1867. In becoming more efficient, explosives also became more lethal.

Growing industry
An Italian manufacturer of gunpowder in his workshop in the 16th century (below).

Time-honoured techniques
Saltpetre (below) and charcoal (far left) are combined to produce rocket-propelled fireworks in Thailand (centre).

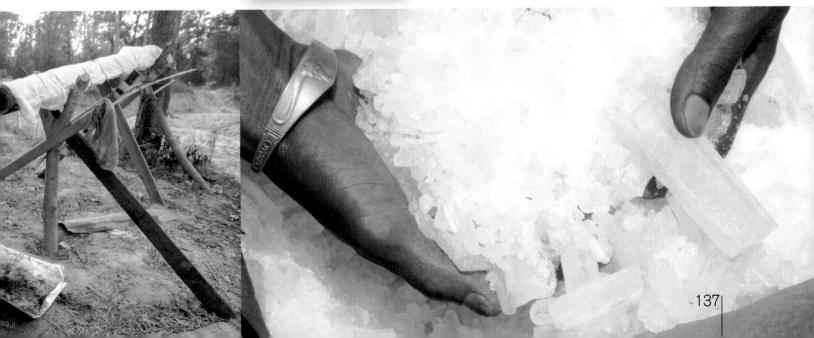

Playing cards C AD900

The oldest known set of playing cards date from the 11th century and were found in Chinese Turkistan. There is evidence, though, that cards were already in use in China a couple of centuries earlier. Some researchers believe that they might have had their origins in a game of dominoes played with tokens printed on cardboard. Cards travelled westward first with the Mongols and then with the Mamluks. They seem to have reached Europe by way of Italy – more precisely, Venice – where they arrived in the luggage of returning traveller Marco Polo. By the end of the 14th century they were known across much of the Continent. One factor that helped to spread them was the growing popularity of woodblock printing, which allowed the cards to be reproduced mechanically.

Games were for the most part tightly controlled and were often subject to government taxes. In 1377 they were forbidden altogether in Florence, while in Paris 20 years later, proclaiming that cards distracted 'artisans and other common people' from their work and families, the provost 'prohibited persons of that class from playing on working days'. The religious authorities were equally ready to see cards as the devil's playthings on the grounds that most games involved gambling and money. But despite the official disapproval, card playing became hugely popular across Europe.

The first game to feature trumps was the English game triumph. Played from 1522, it was the forerunner of whist, from which bridge was devised and eventually standardised in rules by John Collinson in 1885. Poker is derived from *Primero*, a game fashionable at court in Tudor times and played by Henry VIII on the night of his daughter Elizabeth's birth.

Artistic work
Early playing cards, like the example shown left, were individually painted on parchment and were works of art in their own right.

Under suspicion
Painted in 1630, Georges de la Tour's 'The Cheat with the Ace of Clubs' is an early depiction of a card game.

FIGURES AND SYMBOLS

The first sets of cards had staves, cups, swords and coins as the featured suits, as in the packs used for the game of Tarot still popular in the Mediterranean. Diamonds, hearts, spades and clubs were first introduced in France in the late 15th century, along with the court cards representing kings, queens and jacks or knaves. Radical thinkers in France did away with these in the wake of the French Revolution of 1789 in an attempt to remove all symbols of royalty from public and private life, but the banished figures reappeared by popular demand in 1816 when the monarchy was restored. The practice of depicting reversible heads on face cards dates from 1827.

Card pieces
Circular tokens served as playing pieces in an Indian precursor of card games.

The horseshoe c AD900

The hooves of horses in their natural state were originally better adapted to travelling over soft soil than for prolonged galloping over hard ground. Over time, natural evolution combined with the effects of domestication and selective breeding produced bigger, stronger animals with horny hooves. But even these remained too sensitive to cope with the intensive use that humans came to expect of their mounts.

Slip-ons for horses

Some unknown inventor came up with the idea of putting iron under the horse's hooves. The archaeological record remains vague as to when and where this happened – it was possibly in Central Asia, where horse-riding was first practised, or in Greece. It also seems possible that some Germanic peoples were nailing shoes on their horses by the 3rd century BC.

Firmer evidence comes from Latin documents from the early Christian era, which refer to temporary iron shoes held in place by straps, with the sides rising up around the hoof. These 'hipposandals' could be put on the horse for use on hard roads, then removed when no longer required. They were probably used most for draught animals, because of the extra pressure from weight; the underside often had a tread to improve grip. The first nailed shoes known to reach Europe seem to have been introduced from Byzantine lands and their use spread only slowly.

Imperial approval

Horseshoes gained official recognition in the 9th century, when the Byzantine Emperor Leo VI recommended their use. In the West they only became common from the 11th century on, but the farrier or smith soon became a ubiquitous figure in medieval society. The first written advice on the subject dates from the early 16th century; after carefully analysing the anatomy of horses' hooves, the authors recommended adapting the shoes to individual animals. In the same century hot-shoeing, in which horseshoes were heated before fitting, became commonplace. Modern materials allow greater precision in fitting; aluminium is sometimes now chosen for its lightness, and resins may be used to correct imperfections in the hoof before the shoe is put in place.

GOOD-LUCK CHARM

People around the world have come to associate horseshoes with good luck. The most likely explanation for the superstition comes from the almost universal belief, found in India and China as well as Europe, in the power of iron to ward off witches and evil spirits. In traditional societies, a used shoe was often the nearest object made of the metal to come to hand.

Iron hipposandal
Early horseshoes like this one (left) would have been strapped onto the horse's hooves when required.

Celebrating an amazingly modern era

F rom the 3rd century BC on the Chinese showed an inventive genius without parallel in the history of technology and science. Among the innovations they introduced, often many centuries before the West, were street lighting, the parachute, piped drinking water, vaccination, pest control and the magic lantern, the precursor of the cinema.

While Europe anticipated the new millennium with trepidation, the citizens of Canton (now Guangzhou) in southern China were preparing to celebrate the arrival of the Year of the Rat. The New Year festivities saw the whole population out in the streets, watching rockets bursting high up in the sky – a spectacle that would have terrified any visitor from the West, where gunpowder would not be known for another 250 years. In Canton, even the children shrieked with delight.

The firework spectacle unfolded above streets that were lit at night. Methane gas was transported through pipes made of bamboo to lamp-posts erected at intervals of roughly 100m. For the Chinese there was nothing new about street lighting, which had long been installed in big cities across the imperial lands.

The Cantonese had no concerns about the turn of the year 1000, for their calendar was totally different to that of the West. Like the Babylonians, Egyptians and Greeks before them, they knew that the annual round lasted just over 365 days, but the years themselves were counted from the accession of the current emperor. So rather than embarking on a new millennium, the Chinese were entering the second year of the reign of Zhenzong, third ruler of the Song Dynasty. His immediate predecessors had energetically reorganised the administration of the empire and had also taken steps to promote education, a move that would launch a tide of inventions.

High days and holidays

On the main square in front of the governor's palace, the crowd was now cheering a fresh spectacle. An acrobat had just leaped off the top of a 30m-high tower, equipped with

nothing but a giant parasol to cushion his fall. Shaped rather like a pointed hat and perforated with gaps through which air could escape, the device was in effect a parachute; similar contraptions had been known in China for 800 years.

The New Year festivities ran for ten days and were seen by the children as one long holiday. Magic lantern displays were among the treats on offer, with shows advertised every evening. Parents were only too happy to accompany their offspring, for they too enjoyed watching animated images of warriors taking on dragons, huge birds with beating wings, lines of chariots and similar wonders. Their peers had been watching such spectacles for more than a millennium, ever since the lanterns were first invented in about 120 BC. Europeans would wait another 650 years to share such an experience.

In the governor's palace

In his official residence the town's governor received goodwill messages sent by post from provincial officials. Mail services had been available in China for almost 2,000 years, but now mail couriers were able to travel more efficiently by canal, a system of locks having

New Year festival
Just as they do today (background), Chinese people in the Middle Ages (bottom and above right) celebrated the New Year with fireworks, kites, bangers and a host of other entertainments.

recently been installed, again some four centuries before similar ones would make their first appearance in the West. The messengers travelled in boats powered by paddlewheels, which China had first seen five centuries earlier and Europe would adopt five centuries later.

Local dignitaries, immaculately turned out in garments ironed by laundrymaids, sat in respectful attendance on the governor. Irons had been used in China since the 4th century BC – relatively late in comparison with the Egyptians, who had them more than a millennium earlier. The guests gazed admiringly at the decorations in the great hall, especially its phosphorescent paintings, laid down on the putrescent wood of a species of sophora tree, which glittered in the lamplight.

PREDICTING EARTHQUAKES

As early as AD 132 a Chinese inventor named Zang Heng devised the first seismoscope in the form of a large bronze vessel with a ring of eight dragon's heads encircling its rim. When there was seismic activity a ball would emerge from the mouth of a dragon to fall into the gaping mouth of one of the frogs below. The mechanism was triggered by a pendulum that was balanced inside on a central column (right). Sensitive to tiny oscillations, the pendulum triggered a lever that released the ball down one of eight separate tracks.

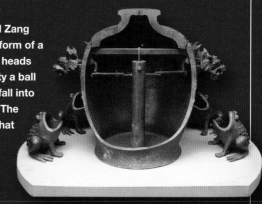

Dinner with a mandarin

From early times the men of letters, scholars and jurists who made up China's mandarin class of administrators and bureaucrats were expected to host dinners on feast days for the people who fell under their jurisdiction. The painting below is a 19th-century depiction of this ancient custom.

Early-warning system

The governor's room also housed a large bronze vase rimmed with eight protruding dragons' heads. Beneath each one sat a frog, gazing upward, open-mouthed. If a bronze ball emerged from one of the dragons' mouths it fell with an ominous clang into the waiting frog below. The noise signalled an earthquake happening somewhere, although it might be some considerable distance away.

The inventor of this ingenious device was Zang Heng, astronomer royal at the Han court, who had first made it (according to our Western calendar) in AD 132. The Canton governor's model was a copy. Almost a thousand years after its creation, Zang's machine was still in demand since earthquakes occurred frequently in China, as they still do today. They brought death, destruction and hardship in their wake, often triggering social unrest (along with demands for emergency supplies of rice) if not open rebellion. Fortunately for the governor, there was to be no seismic activity in that particular year.

A sophisticated infrastructure

The guests that New Year included a number of civil engineers, some of whom were to be rewarded with cash gifts in the form of banknotes – already in circulation for more than a century. One recipient was a man who had upgraded the city's water supply, which featured mechanisms devised more than a millennium before and improved by the hydrologist Bi Lan in the 2nd century.

With running water and street lighting, Canton would have been the envy of Western cities for many centuries to come. London, for example, only got its first street lighting in 1427, when the Mayor, Sir Henry Barton, ordered lanterns to be hung out on winter evenings between Halloween and Candlemas (February 2). As for drinking water, it was first piped into Western homes in London in 1829.

Biological warfare

The peasants were not forgotten at the governor's banquet. The mandarin also had a gift for an individual who had found a new

A REGULAR SUPPLY OF RUNNING WATER

One model of a municipal Chinese water system consisted of a chain of purifying troughs fed by water from a nearby river. The technology was standardised in the reign of Emperor Wen Zong in 828, but improvements continued to be made both for purposes of irrigation and for domestic supplies. As a result, ordinary residents of China's principal cities received water supplies at home, at a time when only the wealthy in Europe had the privilege.

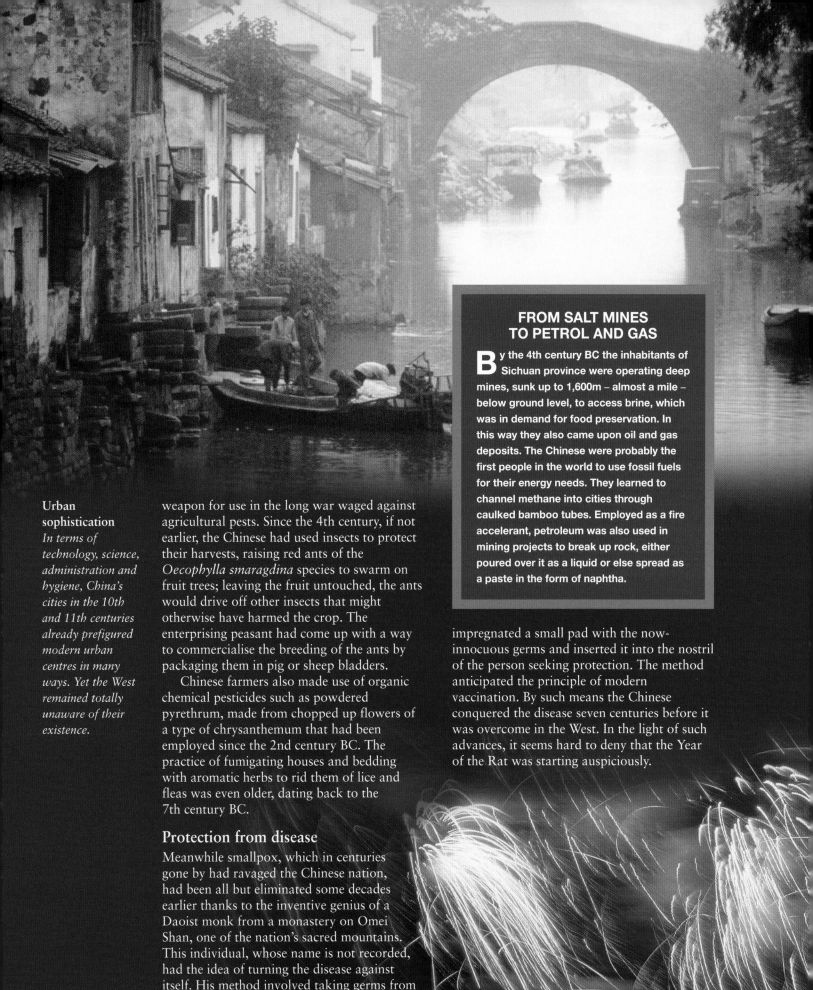

Urban sophistication
In terms of technology, science, administration and hygiene, China's cities in the 10th and 11th centuries already prefigured modern urban centres in many ways. Yet the West remained totally unaware of their existence.

weapon for use in the long war waged against agricultural pests. Since the 4th century, if not earlier, the Chinese had used insects to protect their harvests, raising red ants of the *Oecophylla smaragdina* species to swarm on fruit trees; leaving the fruit untouched, the ants would drive off other insects that might otherwise have harmed the crop. The enterprising peasant had come up with a way to commercialise the breeding of the ants by packaging them in pig or sheep bladders.

Chinese farmers also made use of organic chemical pesticides such as powdered pyrethrum, made from chopped up flowers of a type of chrysanthemum that had been employed since the 2nd century BC. The practice of fumigating houses and bedding with aromatic herbs to rid them of lice and fleas was even older, dating back to the 7th century BC.

Protection from disease

Meanwhile smallpox, which in centuries gone by had ravaged the Chinese nation, had been all but eliminated some decades earlier thanks to the inventive genius of a Daoist monk from a monastery on Omei Shan, one of the nation's sacred mountains. This individual, whose name is not recorded, had the idea of turning the disease against itself. His method involved taking germs from a smallpox victim's pockmarks and inactivating them by heating them. He next

impregnated a small pad with the now-innocuous germs and inserted it into the nostril of the person seeking protection. The method anticipated the principle of modern vaccination. By such means the Chinese conquered the disease seven centuries before it was overcome in the West. In the light of such advances, it seems hard to deny that the Year of the Rat was starting auspiciously.

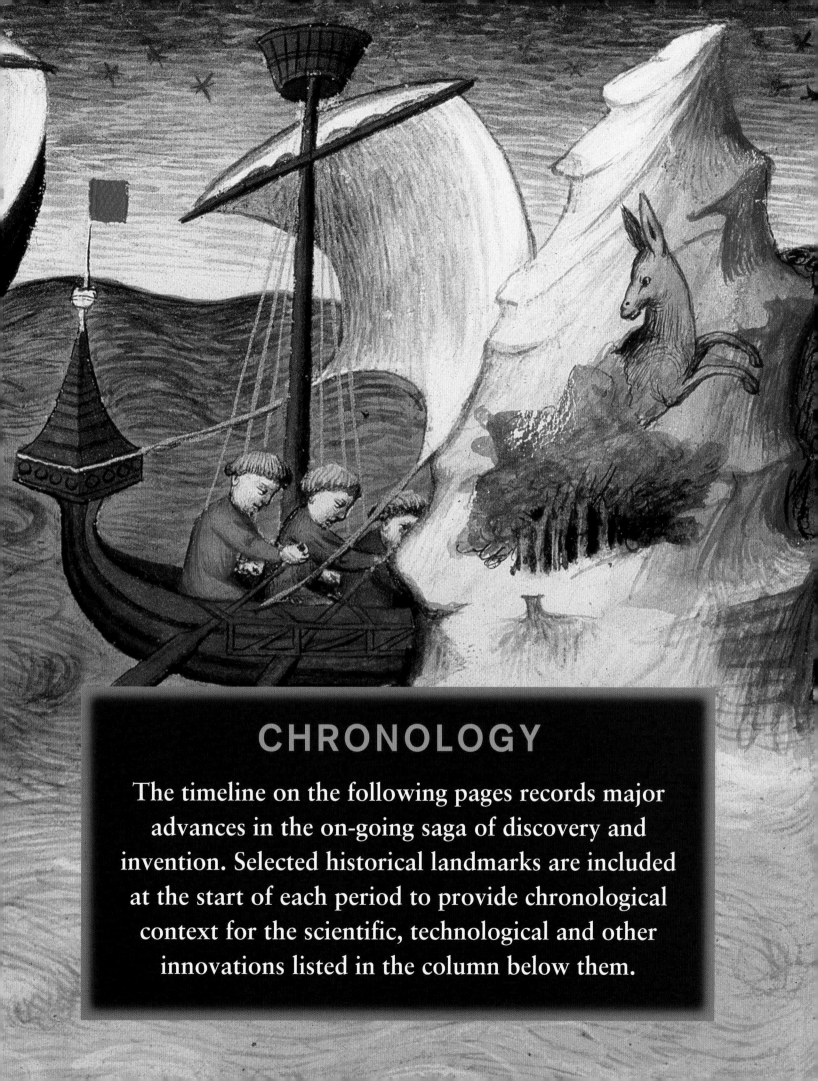

CHRONOLOGY

The timeline on the following pages records major advances in the on-going saga of discovery and invention. Selected historical landmarks are included at the start of each period to provide chronological context for the scientific, technological and other innovations listed in the column below them.

1300 BC	1200 BC	1100 BC	1000 BC
EVENTS			
• The Hittite state declines following the Battle of Kadesh against the Egyptians in 1275 BC; both sides claimed victory • Hebrew exodus from Egypt	• Celtic civilisation enters its golden age • End of the Hittite Empire	• Collapse of the Mycenaean culture of ancient Greece • End of the New Kingdom in Egypt	• The Assyrian Empire in Mesopotamia and the Olmec realm in Mexico are both flourishing • Phoenicians cross the Strait of Gibraltar from Africa to Spain
INVENTIONS			
• Pharaoh Rameses II orders the construction of a canal to link the Red Sea to the Nile and hence to the Mediterranean • The Hittites mobilise a force of 3,500 war chariots against the Egyptians, who were equipped with fewer but lighter, more manoeuvrable chariots • Lacquering techniques for furnishings, statues and crockery are developed in China • The triangular harp is introduced in Egypt	• Iron-working gets under way in Asia Minor • The Phoenicians introduce the first alphabet, composed of 22 letters • Phoenician mariners are using round-hulled boats	• Medical compendia listing healing potions are drawn up in Assyria	• Kites are popular in China, both for pleasure and practical purposes • The Greek alphabet starts to develop, taking the Phoenician alphabet as its starting point • The oldest known map is drawn on papyrus in Egypt

▲ Iron dagger and sheath

◀ Phoenician alphabetic writing

▼ Egyptian map inscribed on papyrus

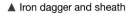

▲ Chinese kites

900 BC

- The first Etruscan cities are founded on the Italian peninsula
- The earliest Greek city-states emerge

- Assyria becomes one of the first states to develop a road network
- Alphabetic scripts appear in the eastern Mediterranean

▲ Byzantine Greek writing

800 BC

- The first Olympic Games are held in 776 BC
- The *Iliad* and the *Odyssey* take shape
- Rome is founded in 753 BC

- Qanats are being dug for water management in Armenia
- Iron saws are in use in Assyrian lands

700 BC

- A Babylonian army sacks Nineveh (612 BC)

- Biremes (galleys with two banks of oars) are introduced in Greece
- Greek infantry detachments adopt the phalange formation
- Etruscans are using false teeth
- The Latin alphabet is in use by the Etruscans
- A lime-based mortar is used in construction in Greece and Mesopotamia
- Metal coins are in circulation in Lydian, the realm of King Croesus; the cities of Ionia and the island of Aegina are the first Greek outposts to adopt the new money

▼ Pheidippides brings news to Athens of the Greek victory at Marathon

600 BC

- The statesman Solon lays the foundations of Athenian democracy from 594 BC onwards

- Thales, a Greek philosopher and mathematician, sets out the five theorems that will make him one of the fathers of geometry
- Eupalinos of Megara constructs a tunnel more than 1km in length through Mount Kastro on the Greek island of Samos, employing early surveying techniques
- The Babylonians adopt a calendar based on the lunar month and the solar year
- Cyrus the Great sets up the world's first postal service in the Persian Empire
- Pythagoras, the Greek philosopher and mathematician, proposes mathematical formulae and theories about the position of the heavenly bodies; calculations involving the Golden Ratio; and principles governing the diffusion and refraction of sound
- Locks employing pins and keys are in use in Egypt

▼ Phoenician galley with multiple banks of oars

▲ Lydian gold coin

500 BC

EVENTS

- The golden age of Athens under Pericles, also known as the 'Age of Pericles' (461–429 BC)
- The Law of the 12 Tables is set down in Rome (450-451BC)

INVENTIONS

- The Greek astronomer Parmenides provides an explanation for the phases of the Moon
- The Greeks invent two separate numerical systems, the Attic and the Ionian
- Rome adopts bronze coinage
- Maps feature in the *Yugong*, the oldest surviving Chinese geographical treatise
- The first catapults – weapons using the principle of leverage – come into service; Archimedes will describe their operation two centuries later
- The oldest surviving knotted carpet, found on the China–Mongolia border, dates from this time
- Herodotus travels widely to research his *Histories*, which will win him the titles of Father of Geography and Father of History
- Hippocrates rationalises Greek medicine, bringing together all current knowledge of anatomy and physiology
- The Carthaginian mariner Hanno explores the west coast of Africa

400 BC

EVENTS

- The Grand Canal is built in China
- The city of Alexandria is founded in Egypt by Alexander the Great (332 BC)

INVENTIONS

- Greek engineers produce improved ballistic weapons, including ballistae and catapults
- Irons resembling brass warming-pans are used in China to press laundry; the devices are filled with hot embers or charcoal
- Pulleys are invented, reputedly by Archytas of Tarentum, who is also credited with the water-raising machine that in time becomes known as Archimedes' screw
- Saddles are in use by Central Asian horsemen
- In China, blast furnaces are used to produce cast iron
- The first chemical weapons employing toxic gases are deployed in China
- Romans start to build the vast road network that will eventually criss-cross the empire; work commences on the Via Appia
- The Greek explorer Pythias sails to the British Isles and beyond
- The abacus is in use in Asia

300 BC

EVENTS

- Emperor Asoka rules the Mauryan Empire in India (273–232 BC)

INVENTIONS

- Greek engineers devise gears for use in devices such as water clocks and the Archimedes' screw
- Pharaoh Ptolemy I Soter makes Alexandria in Egypt the foremost centre of learning of its day, founding the city's celebrated library
- The Greek astronomer Aristarchus of Samos challenges conventional wisdom by asserting that the Earth revolves around the Sun; he also calculates the diameter of the Moon
- Sostratos of Cnidos builds the Lighthouse of Alexandria, one of the Seven Wonders of the Ancient World
- The Greek engineer and mathematician Archimedes writes seminal works that will shape the future course of physics and mathematics; he also devises practical uses for the principle of leverage, which he had earlier defined mathematically
- Philo of Byzantium first describes the principle of chain transmission

▲ Roman mile-marker

A box iron ▶

◀ Wooden pulley

▲ The Lighthouse of Alexandria

300 BC

- First Punic War between Rome and Carthage (265–241 BC)
- Qin Shihuangdi becomes First Emperor of China (221–210 BC)
- The Roman general Scipio Africanus conquers the Iberian peninsula (206 BC)

- New devices based on the principle of the Archimedes' screw are brought into operation to draw water

- The Greek engineer and scientist Ctesibius, one of the founders of the School of Alexandria, is credited with the invention of the suction pump, compressed air weapons, the hydraulis or water organ and an astronomical clock

- The Greek mathematician Euclid writes the *Elements*; the system he describes, dubbed Euclidian geometry, will remain unchallenged until modern times – one well-known axiom, accepted to this day, maintains that the number of prime numbers is infinite

- China's political unification entails the development of an imperial road network

- The first relief map is drawn in China

- The Brahmi numerical system, based on the power of nine, comes into use in India

- Roman engineers learn how to construct arches, bringing about a major advance in bridge-building techniques

- The Greek scientist Eratosthenes – mathematician, astronomer, geographer, poet and director of the library of Alexandria – becomes the first person to calculate the circumference of the Earth

▲ An angel fancifully employs a parachute

◄ The hydraulis, an ancestor of the organ

▲ Archimedes puts a theory to the test

◄ Archimedes' screw

200 BC

- Patanjali writes the *Yoga Sutras* (c150 BC)
- Carthage is destroyed by Rome (146 BC)
- The Romans take control of Greece (146 BC)

- Greek farmers use wine and oil-presses, powered by a perpetual-screw mechanism

- Chinese acrobats perform the world's first parachute jumps

- The gimbal mechanism is developed in China for burning incense

- The Greek astronomer Hipparchus, one of the principal luminaries of the School of Alexandria, draws up a catalogue of more than 1,000 stars

- Magic lanterns are used in China to project images

- Egyptian farmers employ the sakia, a new version of the noria water-wheel; the breakthrough development is to use gearing to transfer energy from a horizontal wheel turned by a draft animal to a vertical wheel that raises water

- Parchment is invented at Pergamon in Asia Minor

- Greek astronomers use astrolabes to measure the inclination of stars in the night sky

Indian astrolabe ►

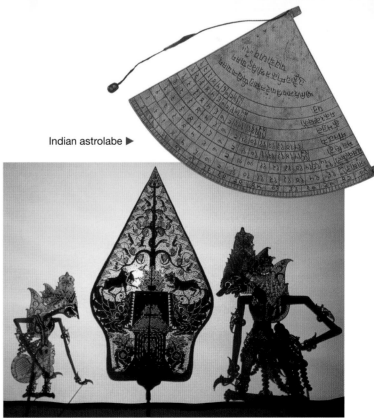

▲ Indonesian shadow theatre

100 BC   **0**   **AD 100**

EVENTS

- Cleopatra VII is queen of Egypt (51–30 BC)
- The Julian calendar is adopted in Rome (46 BC)
- Start of Ptolemaic Egypt (30 BC)

- Xin Dynasty in power in China (AD 9–23)
- St Paul spreads the Christian message (cAD 50)
- The Colosseum is built in Rome (AD 72)
- The Romans conquer Britain (AD 43)

- The Pantheon is built in Rome (118–120)
- The Bar Kochba Revolt breaks out in Israel (132–135)

INVENTIONS

- Hero of Alexandria puts his name to several inventions including lifting devices employing pulleys, the dioptra (an early theodolite), the first working steam engine and various automata

- An unknown Chinese state official introduces paper as a medium for writing

- Chemical pesticides are used in China

- Greek mariners use star charts for navigation

- Roman engineers and construction workers now have available a water-resistant form of concrete and mortar that enables them to build vast bridges and aqueducts

- The Roman engineer and architect Vitruvius describes the working principle of the crane

- Varron's 41-volume *Antiquities* is the world's first encyclopaedia

- Vitruvius explains the mechanism of lifts and how a water-mill works

- Emperor Augustus establishes a postal service in the Roman Empire known as the *cursus publicus*

- Heavy iron ploughs come into use in the Roman world

- The houses of wealthy Romans set new standards of hygiene and comfort, with running water and hypocausts (underfloor central heating)

- Roman artisans develop a technique for making stained glass

- The wheelbarrow makes its first appearance in China

- The Greek polymath Ptolemy, the foremost astronomer, geographer and mathematician of his day, writes the *Almagest* in Alexandria; in its pages he proposes a mathematical model for working out the apparent motions of the planets

- Scissors are in use by Romans

- The first seismograph is produced in China

- Chinese mariners sailing junks with squared-off hulls and lateen rigging reach the coast of India

A model of the first steam engine devised by Hero of Alexandria ▶

▼ A scale model of a water-wheel

▼ Paper – a new medium for the written word

AD 200

- Ardashir I founds Persia's Sassanid Empire (200)
- Fall of China's Han Dynasty (220)

- Iron skates are in use in Scandinavia

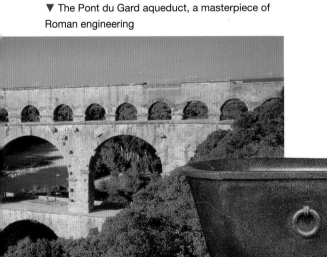

▼ The Pont du Gard aqueduct, a masterpiece of Roman engineering

AD 300

- Constantinople is founded (324)

- An elaborate water-mill complex is built at Barbegal in southern France
  - The Chinese invent stirrups, giving riders a more secure seat on their horses

▲ Primitive lift

AD 400

- St Augustine writes the *Confessions* (c400)
- The Romans leave Britain (410)
- Fall of the Western Roman Empire (476)

- Indian mathematicians introduce the concept of zero and incorporate it into their decimal numerical system

- Roman legionaries adopt the long sword known as the spatha

- Bark paper is in use in Mexico

▲ Leaded windows, made using stained-glass techniques

AD 500

- The Byzantine Emperor Justinian introduces his legal code (534)
- Buddhism reaches Japan (552)
- Columba founds his monastery on Iona (563)

- The first inscriptions employing an alphabetic Arabic script date from this time

- A Byzantine scholar named John Philoponus describes a planispheric astrolabe

- In China and Japan stencils and stone seals are used to reproduce texts

▼ Zero – a universal symbol

▲ Chinese wheelbarrow

◄ A Roman bathtub made of porphyry

151

AD **600**

AD **700**

AD **800**

EVENTS

- The Hijra, Muhammad's migration from Mecca to Medina, marks the start of the Muslim calendar (622)
- The Mayan temples are built at Tikal (680–700)

- Baghdad is founded as the capital of the Abbasid Dynasty (762)

- The first Maoris arrive in New Zealand (c800)
- Charlemagne is the first Holy Roman Emperor (800–814)
- The House of Wisdom is founded in Baghdad (c832)
- St Cyril and St Methodius begin converting the Slavs (863)
- The Vikings land in Ireland (c875)

INVENTIONS

- An Indian Brahman invents a game of strategy – chess

- Isidore of Seville becomes the first chronicler to mention the use of a goose quill as a writing implement

- The world's first recorded skyscraper is built in China – a cast-iron pagoda 90m high

- The Byzantines perfect the composition of the incendiary Greek fire

- Woodblock printing is introduced in China

- The Vikings undertake long-distance voyages across the Atlantic Ocean

- Hucbald of St Amand, a Benedictine monk, writes the first work on Western musical theory

- Gunpowder is invented in China

▲ A knight from the Lewis chess set

▲ Tibetan woodblock print

▲ Neumes – the earliest form of musical notation

▼ Byzantine naval forces using Greek fire

AD 900

- The feudal system emerges in Europe
- The Benedictine abbey of Cluny is founded in France (910)
- Toltecs gain ascendancy in Mexico (c900)
- Start of the Song Dynasty in China (960)

- Playing cards are first used in China

- The Byzantine Emperor Leo VI recommends the use of horseshoes

- A woodblock edition of Confucian classics is printed in China

- Gerbert d'Aurillac introduces Arabic numerals and the astrolabe to Europe

- Vaccination against smallpox is introduced in China

- Camshafts are added to watermills to drive saws, bellows and blacksmiths' fullering tools

- The Viking explorer Eric the Red discovers Greenland

AD 1000

- Confucianism becomes China's official state ideology (1020)
- Norman invaders conquer England (1066)
- The First Crusade establishes a Christian state in the Holy Land (1095–1099)
- The Domesday Book is compiled in England (1083)

- The horse collar becomes standard equipment for draught animals in northern Europe

- The Persian physician and philosopher Ibn Sina (Avicenna) writes his *Canon of Medicine,* summing up the medical knowledge of his day

- Guido d'Arezzo, a Benedictine monk, introduces the stave into musical notation

- Al-Biruni compiles his 11-volume *Masudi Canon,* a comprehensive encyclopaedia of scientific knowledge

- A Chinese military work, the *Wujing Zongyao,* makes the first recorded reference to a magnetic compass; the device will not reach Europe for another century

- Forks are introduced in Venice

- Terracotta moveable-type characters are used in printing in China

AD 1100

- The medieval code of chivalry emerges in Europe
- The Knights Templar military order is founded (c1119)
- Saladin conquers Egypt (1171–1193)

- The use of heavy wheeled ploughs equipped with ploughshares spreads across northwestern Europe, boosting agricultural productivity

- Windmills come into use in Europe, with more advanced sails than their Persian predecessors

- Stern-post rudders, known in China since the 1st century, begin to replace side-rudders in European shipping

- The pointed arch makes its first appearance in the Basilica of St Denis, near Paris, marking the start of the Gothic style of architecture

◀ Horseshoe

▲ Parchment playing card

▼ Page from an Arabic encyclopaedia

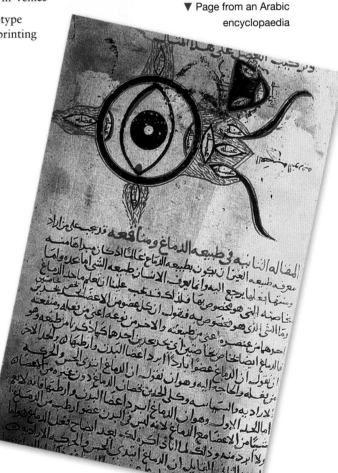

Index

Page numbers in *italics* refer to captions.

Picture credits

THE ADVENTURE OF DISCOVERIES AND INVENTIONS
Iron Age to Dark Age – 1200BC to AD1000
is published by The Reader's Digest Association Limited,
11 Westferry Circus, Canary Wharf, London E14 4HE

Copyright © 2009 The Reader's Digest Association Limited

The book was translated and adapted from *De L'Âge du Fer à l'Âge des Ténèbres*, part of a series entitled L'ÉPOPÉE DES DÉCOUVERTES ET DES INVENTIONS, created in France by BOOKMAKER and first published by Sélection du Reader's Digest, Paris, in 2005.

Translated from French by Tony Allan

Series editor Christine Noble
Art editor Julie Bennett
Designer Martin Bennett
Consultant Ruth Binney
Proofreader Ron Pankhurst
Indexer Marie Lorimer

Colour origination Colour Systems Ltd, London
Printed and bound in China

READER'S DIGEST GENERAL BOOKS
Editorial director Julian Browne
Art director Anne-Marie Bulat
Managing editor Nina Hathway
Head of book development Sarah Bloxham
Picture resource manager Christine Hinze
Pre-press account manager Dean Russell
Product production manager Claudette Bramble
Production controller Sandra Fuller

Copyright © 2009 The Reader's Digest Association Far East Limited
Philippines Copyright © 2009 The Reader's Digest Association Far East Limited
Copyright © 2009 The Reader's Digest (Australia) Pty Limited
Copyright © 2009 The Reader's Digest India Pvt Limited
Copyright © 2009 The Reader's Digest Asia Pvt Limited

We are committed to both the quality of our products and the service we provide to our customers. We value your comments, so please feel free to contact us on 08705 113366 or via our website at **www.readersdigest.co.uk**

If you have any comments or suggestions about the content of our books, you can email us at **gbeditorial@readersdigest.co.uk**

CONCEPT CODE: FR0104/IC/S
BOOK CODE: 642-002 UP0000-1
ISBN: 978-0-276-44514-9
ORACLE CODE: 356400002H.00.24